Applied Pisciculture

ISBN: 978-93-6726-471-3

Author

Dr. Manoj Kandpal
Prof. Ashok Kumar
Mrs.Trupti Dhananjay Kakulte
Mrs.Vaishali Ashok Rahane
Mrs.Sonali Ramesh Deore

BOOK SAGA®
PUBLICATIONS

Title: Applied Pisciculture

ISBN: 978-93-6726-471-3

Author: Dr. Manoj Kandpal, Prof. Ashok Kumar, Mrs.Trupti Dhananjay Kakulte, Mrs.Vaishali Ashok Rahane, Mrs.Sonali Ramesh Deore

Published by: Book Saga Publications
Rahuri Factory, Rahuri, Ahmednagar, Maharashtra, India, 413706.
Website - www.booksagapublications.com
E-mail - contact@booksagapublications.com
Mobile- +918788403871
First Published - October, 2025

Contents

Abstract

The book *"Applied Pisciculture"* delves into the scientific principles and practical applications that form the foundation of modern fish farming and aquaculture. It integrates biological understanding, technological innovation, and management strategies to present a complete overview of fish culture as a sustainable and profitable enterprise. The objective of the book is to bridge the gap between theoretical pisciculture and field-level practice by focusing on applied methodologies that enhance productivity, efficiency, and environmental harmony.

The content emphasizes the life cycle management of cultivable fishes, hatchery operations, water and pond management systems, feed formulation, and disease prevention techniques. Each chapter highlights not only the technical details but also the economic, ecological, and social relevance of aquaculture in today's food systems. With increasing global attention on sustainable protein production, *Applied Pisciculture* underscores the importance of responsible fish farming that ensures high yield while maintaining ecological balance.

This book serves as a comprehensive resource for students, academicians, fishery professionals, and entrepreneurs engaged in aquaculture. It combines core scientific knowledge with modern advancements such as integrated aquaculture systems, biosecurity protocols, and smart aquaculture technologies. Ultimately, *Applied Pisciculture* seeks to empower readers with both conceptual clarity and hands-on understanding to promote innovation, sustainability, and profitability in fish culture practices across diverse aquatic ecosystems.

Preface

The evolution of fish culture from a traditional subsistence activity to a highly organized scientific enterprise represents one of the most remarkable transitions in agricultural history. *Applied Pisciculture* has been written with the vision of presenting this transformation through a structured and comprehensive approach that captures both the scientific essence and practical applicability of fish farming.

This book is a culmination of careful study, field observation, and synthesis of established and emerging trends in aquaculture. Each chapter is designed to serve as a standalone learning unit, providing in-depth knowledge on breeding, culture systems, pond management, nutrition, health, and economic evaluation. The intention is to present pisciculture not merely as a biological discipline, but as an applied science that interacts with technology, economics, and environmental conservation.

A major focus has been placed on sustainability, innovation, and adaptability — the three pillars that define modern pisciculture. From seed production to market readiness, every stage of the aquaculture process has been discussed in a way that enables students and practitioners to adopt best practices suited to varying geographical and ecological conditions. The chapters also reflect the integration of traditional wisdom with contemporary technology, thus offering a balanced perspective for both developing and advanced aquaculture regions.

It is hoped that *Applied Pisciculture* will inspire readers to explore, innovate, and contribute meaningfully to the ever-expanding world of fish farming. Whether as a textbook, a professional guide, or a reference for entrepreneurs, this work is intended to serve as a lasting contribution to the advancement of applied fish culture.

Acknowledgement

The completion of *Applied Pisciculture* would not have been possible without the collective inspiration, guidance, and support received from several individuals and institutions.

First and foremost, sincere gratitude is extended to the academic mentors and fisheries experts whose knowledge and encouragement have shaped the author's understanding of aquaculture. Their unwavering dedication to research, education, and sustainable fisheries management has been a constant source of motivation.

Special thanks are due to colleagues and technical experts in the field of aquaculture who provided valuable insights, shared data, and reviewed technical sections with keen attention to detail. Their expertise ensured that the book remained scientifically sound and practically relevant.

Heartfelt appreciation is also expressed to the research laboratories, aquaculture training centers, and fish breeding units whose resources and facilities greatly contributed to the compilation of field-based examples and case studies included in this book.

A note of deep gratitude is extended to friends and family members for their constant patience, encouragement, and understanding throughout the writing process. Their faith and moral support have been invaluable.

Finally, the author acknowledges all readers, students, and professionals who continue to contribute to the field of pisciculture. Their curiosity, innovation, and passion for sustainable aquaculture remain the driving force behind this endeavor.

CHAPTER 1

INTRODUCTION TO APPLIED PISCICULTURE

Key Outcomes

Upon completing this chapter, readers will be able to:

1. Define pisciculture and articulate its scope within the broader context of aquaculture and food production systems
2. Trace the historical development of fish culture from ancient practices to modern commercial operations
3. Explain the significance of pisciculture in global food security, nutritional supplementation, and economic development
4. Distinguish between extensive, semi-intensive, and intensive fish culture systems based on their operational characteristics
5. Identify and apply fundamental principles governing successful applied fish farming operations
6. Analyze current global and regional trends shaping the future trajectory of pisciculture development

Abstract

This introductory chapter establishes the foundational framework for understanding applied pisciculture as a scientific discipline and commercial practice. It explores the definition, scope, and historical evolution of fish culture, tracing its development from ancient subsistence practices to modern intensive aquaculture systems. The chapter examines the critical role of pisciculture in addressing global food security challenges, contributing to economic development, and providing livelihood opportunities across diverse geographical regions. Different culture systems—extensive, semi-intensive, and intensive—are systematically described, highlighting their operational characteristics, resource requirements, and production capabilities. The fundamental principles governing successful fish farming operations are articulated, emphasizing biological compatibility, environmental sustainability, and economic viability. Finally, the chapter analyzes contemporary global and regional trends in pisciculture development, identifying emerging opportunities, technological innovations, and challenges facing the aquaculture sector. This comprehensive introduction prepares readers for deeper exploration of specific technical aspects covered in subsequent chapters.

1.1 Definition and Scope of Pisciculture

Pisciculture represents the scientific art and practice of breeding, rearing, and harvesting fish under controlled or semi-controlled environmental conditions for human consumption, ornamental purposes, sport fishing, or conservation objectives. The term originates from the Latin words *piscis*, meaning fish, and *cultura*, meaning cultivation or rearing. In its broadest interpretation, pisciculture encompasses all aspects of fish production, from selection of suitable species and breeding stock management to providing optimal growing conditions, nutrition, health care, and ultimately harvesting marketable products.

The scope of pisciculture extends far beyond simple fish rearing. It constitutes an interdisciplinary field integrating knowledge from biology, ecology, genetics, nutrition, engineering, economics, and management sciences. Modern pisciculture practitioners must understand fish anatomy, physiology, and behavior to create suitable culture environments. They must comprehend ecological principles to maintain water quality and manage disease organisms. Knowledge of genetics enables selective breeding for improved growth rates, disease resistance, and market characteristics. Understanding nutrition science allows formulation of cost-effective feeds that optimize growth and health. Engineering knowledge facilitates design of efficient culture systems, water management infrastructure, and harvesting equipment. Economic analysis guides investment decisions, production planning, and marketing strategies.

Applied pisciculture specifically emphasizes the practical implementation of scientific knowledge in commercial fish production operations. Unlike theoretical ichthyology or pure fish biology, applied pisciculture focuses on translating research findings into field-applicable technologies and management practices that enhance production efficiency, profitability, and sustainability. This applied orientation distinguishes pisciculture textbooks and training programs from purely academic studies of fish biology.

The geographical scope of pisciculture spans tropical, subtropical, and temperate regions, with practices adapted to local environmental conditions, available resources, cultural preferences, and market demands. Tropical pisciculture often emphasizes warm-water species such as carps, tilapias, and catfishes cultured in earthen ponds. Temperate regions may focus on salmonids, trouts, and cold-water species reared in raceways, tanks, or cage systems. Coastal and marine pisciculture includes cultivation of sea bass, groupers, pompanos, and other brackish or saltwater species.

Species diversity within pisciculture encompasses hundreds of cultivable fish species worldwide. Major groups include freshwater carps forming the backbone of Asian aquaculture, African catfishes and tilapias providing protein across the developing world, salmonids serving premium markets in developed nations, and ornamental fishes satisfying the global aquarium trade. Each species group presents unique biological requirements, culture protocols, and market characteristics that practitioners must understand and accommodate.

Production objectives in pisciculture vary considerably depending on target markets and social contexts. **Food fish production** dominates commercial pisciculture, providing affordable animal protein to billions of consumers globally. **Ornamental fish culture** supplies decorative species for home and public aquaria, generating substantial economic value from relatively small-scale operations. **Sport fish production** supports recreational fishing industries by stocking public and private waters. **Conservation aquaculture** breeds endangered species for population restoration and genetic preservation. Each objective requires specialized knowledge, infrastructure, and management approaches.

The economic scope of pisciculture has expanded dramatically in recent decades. Global aquaculture, dominated by fish production, now supplies over half of all fish consumed by humans, surpassing capture fisheries in total output. This rapid growth reflects increasing demand for fish protein, stagnating or declining wild fish stocks, and successful development of efficient culture technologies. Pisciculture operations range from small subsistence farms producing a few hundred kilograms annually to industrial-scale facilities yielding thousands of tons. The sector employs millions of people worldwide in direct production, feed manufacturing, equipment supply, processing, distribution, and related services.

Environmental considerations increasingly shape the scope and practice of modern pisciculture. Sustainable production methods minimize negative environmental impacts while maintaining economic viability. **Integrated culture systems** combine fish production with crop agriculture or livestock rearing, recycling nutrients and reducing waste discharge. **Recirculating aquaculture systems** treat and reuse water, enabling high-density production with minimal environmental footprint. **Ecosystem-based management** approaches consider broader environmental impacts and benefits of aquaculture operations. Growing consumer awareness and regulatory frameworks increasingly demand environmentally responsible production practices.

Social and cultural dimensions also define pisciculture's scope. In many developing regions, small-scale fish farming provides household nutrition, supplementary income, and rural employment opportunities. Cultural preferences strongly influence species selection, product forms, and marketing strategies. Religious and ethnic traditions may dictate acceptable species, processing methods, and consumption patterns. Gender roles in fish production and marketing vary across cultures, affecting technology adoption and benefit distribution. Understanding these social contexts remains essential for successful pisciculture development.

Technological scope continues expanding as innovations enhance production efficiency and sustainability. **Genetic improvement programs** develop faster-growing, disease-resistant strains. **Precision aquaculture** employs sensors, automation, and data analytics to optimize feeding, water quality management, and health monitoring. **Biotechnology applications** include hormone-based breeding induction, probiotics for disease prevention, and biofloc technology for water quality management. **Digital platforms** facilitate market linkages, technical information dissemination, and supply chain coordination.

1.2 Historical Development of Fish Culture

The practice of fish culture represents one of humanity's oldest forms of animal husbandry, with archaeological and written evidence documenting fish rearing activities spanning several millennia. Understanding this historical progression provides valuable perspective on how traditional practices evolved into modern scientific aquaculture systems and reveals enduring principles that continue guiding contemporary operations.

Ancient Origins and Early Practices

The earliest documented fish culture activities appear in ancient Chinese texts dating to approximately 2000 BCE, describing carp rearing in natural and artificial ponds. The Chinese treatise attributed to Fan Li, written around 475 BCE, provides detailed guidance on pond construction, species selection, stocking densities, and management practices remarkably similar to principles still followed today. These ancient Chinese practitioners recognized that common carp, grass carp, silver carp, and other species occupied different ecological niches within pond ecosystems, enabling **polyculture systems** that efficiently utilized available natural food resources.

Ancient Egyptian tomb paintings and hieroglyphics from approximately 2500 BCE depict fish held in decorative ponds, suggesting both ornamental and food production purposes. Roman civilization extensively practiced fish culture, constructing elaborate systems of coastal ponds and inland facilities for rearing various species. Wealthy Roman citizens maintained private piscinas stocked with prized species, while commercial operations supplied urban markets. Roman engineers developed sophisticated water management systems including channels, gates, and filtration methods that influenced European aquaculture for centuries afterward.

Medieval European monasteries preserved and advanced fish culture knowledge during periods when classical learning declined. Monks raised carp, trout, pike, and other species in purpose-built ponds to provide protein during religious fasting periods when meat consumption was prohibited. Monastic aquaculture manuals documented seasonal management practices, feeding techniques, and simple breeding methods. These institutions established the foundation for European freshwater fish culture that persisted into the modern era.

Hawaiian ancient aquaculture represented another independent development of sophisticated fish culture systems. Native Hawaiians constructed over 400 coastal fishponds using walls of coral and lava rock to create enclosed brackish-water environments. These ponds cultured mullet, milkfish, and other species, integrating aquaculture with broader resource management systems that sustained substantial human populations on isolated Pacific islands. The engineering skill and ecological understanding demonstrated in these traditional systems commands respect even by contemporary standards.

Development of Scientific Aquaculture

The transformation of traditional fish culture into modern scientific aquaculture began during the 18th and 19th centuries in Europe. German and French scientists systematically studied fish reproduction, developing techniques for **artificial propagation** through manual stripping of eggs and milt, external fertilization, and incubation in controlled environments. These breakthrough discoveries enabled reliable seed production independent of natural spawning conditions, dramatically expanding culture possibilities.

Stephen Ludvig Jacobi, an 18th-century German fish culturist, pioneered artificial fertilization methods for trout, demonstrating successful hatching of eggs fertilized outside the fish body. French scientists Joseph Remy and Antoine Géhin refined these techniques in the 1840s, establishing the first successful trout hatchery operation. Their work proved that fish reproduction could be controlled and manipulated, laying groundwork for modern hatchery technology.

The late 19th century witnessed rapid expansion of government-sponsored fish culture programs in Europe and North America, primarily focused on stocking declining wild populations. Hatcheries producing salmonids, particularly rainbow trout and Atlantic salmon, proliferated across temperate regions. These operations aimed at conservation and enhancement of sport fisheries rather than commercial food production, but developed fundamental technologies and management practices applicable to commercial aquaculture.

Asian fish culture traditions merged with Western scientific approaches during the colonial period. British administrators in India studied indigenous carp culture practices while introducing European species and management concepts. This cross-fertilization of knowledge systems enriched both traditions. Indian scientist H.C. Chaudhuri and colleagues achieved a major breakthrough in the 1950s by developing **induced breeding techniques** using pituitary gland extracts to stimulate spawning in major carps. This innovation revolutionized Asian carp culture, enabling reliable seed production and supporting massive expansion of freshwater aquaculture across South and Southeast Asia.

Twentieth Century Expansion

The mid-20th century marked accelerating development and commercialization of fish culture worldwide. Post-World War II food security concerns motivated governments to invest in aquaculture research and development. International organizations including the Food and Agriculture Organization of the United Nations promoted aquaculture as a strategy for increasing protein supply in developing nations. Technical assistance programs transferred knowledge and technologies across regions, accelerating adoption of improved practices.

Chinese aquaculture experienced dramatic expansion following socialist reforms emphasizing rural development and food self-sufficiency. The government promoted pond construction, polyculture systems, and integrated farming approaches combining fish culture with agriculture and livestock production. These efforts established China as the world's dominant aquaculture producer, a position maintained to the present day.

Technological advances in the latter 20th century transformed aquaculture from extensive traditional practices toward increasingly intensive commercial systems. Development of nutritionally complete formulated feeds eliminated dependence on natural productivity, enabling higher stocking densities and faster growth rates. Improved understanding of water quality requirements led to development of aeration systems, mechanical filtration, and biofilters supporting intensive culture. Disease management advanced through development of vaccines, therapeutants, and biosecurity protocols. Genetic selection programs produced improved strains exhibiting superior growth, survival, and market characteristics.

Salmon farming emerged as a major global industry during the 1970s and 1980s, pioneered in Norway and subsequently expanding to Chile, Scotland, Canada, and other temperate coastal regions. Floating net cage systems enabled efficient utilization of sheltered marine environments, while selective breeding and nutritional advances dramatically improved production efficiency. Salmon farming demonstrated that aquaculture could supply premium-quality products to sophisticated markets, challenging the perception that cultured fish were inferior to wild-caught products.

Shrimp farming experienced parallel expansion in tropical coastal regions, becoming a major export industry for countries in Asia, Latin America, and elsewhere. Despite environmental controversies and disease challenges, shrimp aquaculture illustrated the substantial economic value intensive aquaculture could generate.

Contemporary Era and Future Directions

The 21st century has witnessed continued rapid growth and technological sophistication in global pisciculture. Aquaculture now provides approximately 50 percent of all fish consumed by humans, with the proportion continuing to increase as wild capture fisheries stagnate or decline. Production has shifted toward increasingly intensive systems employing advanced technologies for environmental control, health management, and production optimization.

Recirculating aquaculture systems represent a major contemporary innovation, employing sophisticated water treatment technologies to enable high-density fish production with minimal water consumption and waste discharge. These systems can be established virtually anywhere, including urban and desert locations far from traditional water resources. They offer biosecurity advantages, environmental benefits, and potential for producing premium-quality products commanding price premiums.

Genetic improvement continues advancing rapidly, with selective breeding programs producing superior strains of major species. Molecular genetics technologies enable marker-assisted selection, dramatically accelerating genetic gain. Genetically improved farmed tilapia, selectively bred carps, and domesticated salmon strains exhibit substantially enhanced production characteristics compared to wild progenitors.

Integrated multi-trophic aquaculture represents another contemporary innovation, cultivating multiple species occupying different trophic levels in interconnected systems. For example, salmon or fish effluents provide nutrients for seaweed and shellfish production, creating integrated systems mimicking natural ecosystem functions while reducing environmental impacts.

Digital technologies and precision aquaculture employ sensors, automation, artificial intelligence, and big data analytics to optimize production processes. Automated feeding systems respond to fish appetite, reducing waste and improving feed conversion. Sensor arrays continuously monitor water quality parameters, alerting managers to problems before fish health deteriorates. Computer vision systems assess fish growth, health, and behavior, enabling early disease detection and optimal harvest timing.

Despite remarkable progress, pisciculture faces ongoing challenges including disease management, environmental sustainability, climate change impacts, market development, and social acceptance. Future development will likely emphasize sustainable intensification, producing more fish from fewer resources with reduced environmental footprint. Continued technological innovation, improved management practices, and supportive policy frameworks will determine whether pisciculture realizes its potential as a major global food system.

1.3 Importance of Pisciculture in Food Security and Economy

Pisciculture occupies an increasingly critical position in global strategies addressing food security, nutritional adequacy, economic development, and sustainable resource management. Understanding these multifaceted contributions provides essential context for appreciating why investments in aquaculture research, development, and implementation merit high priority.

Contribution to Global Food Security

Food security exists when all people, at all times, have physical and economic access to sufficient, safe, and nutritious food meeting their dietary needs and food preferences for an active and healthy life. Pisciculture contributes fundamentally to all dimensions of food security: availability, access, utilization, and stability.

Fish constitute the most important source of animal protein for billions of people globally, particularly in developing nations where terrestrial livestock may be less accessible or culturally acceptable. Approximately three billion people depend on fish for at least 20 percent of their animal protein intake. In many coastal and island nations, fish provide over 50 percent of dietary animal protein. Pisciculture expansion directly increases food availability in regions where wild fish stocks cannot meet growing demand.

Global fish consumption has increased dramatically over recent decades, driven by population growth, rising incomes, urbanization, and dietary diversification. Per capita fish consumption now exceeds 20 kilograms annually on average worldwide, with much higher levels in many Asian, island, and coastal populations. Without aquaculture expansion, this growing demand could not be met. Wild capture fisheries peaked in the 1990s and show limited scope for further expansion. Indeed, many wild stocks are fully exploited, overexploited, or depleted. Pisciculture now supplies approximately 55 percent of fish for human consumption, with the proportion continuing to increase. This shift from capture to culture represents one of the most significant transformations in global food systems.

Pisciculture offers particular advantages for increasing food production compared to terrestrial animal agriculture. Fish generally exhibit superior **feed conversion efficiency**, requiring less feed to produce a unit of body mass compared to poultry, pigs, or cattle. This efficiency reflects fish being cold-blooded organisms not expending energy maintaining body temperature, and aquatic life supporting their body weight without energy-intensive skeletal structures. Many fish species also occupy lower trophic levels, feeding on plants or plankton rather than requiring animal protein inputs. These biological efficiencies mean aquaculture can produce animal protein with fewer resource inputs per unit output compared to many terrestrial alternatives.

Aquaculture systems can be established in diverse environments including regions unsuitable for conventional agriculture. Marginal lands with poor soil, areas with abundant water resources, coastal zones, and even urban settings can support fish production. This geographical flexibility enables food production in areas unable to contribute significantly to terrestrial agriculture, expanding total productive capacity.

The relatively short production cycle of many cultured fish species enables rapid response to food security needs. Species like tilapia and catfish reach market size in six to twelve months, compared to years for cattle or even longer for tree crops. This quick turnover enables rapid production scaling when demand increases, providing resilience against food supply disruptions.

Nutritional Contributions

Beyond simply providing protein and calories, fish supply essential nutrients often deficient in populations heavily dependent on plant-based diets. Fish provide high-quality **complete protein** containing all essential amino acids in appropriate proportions for human nutrition. This protein exhibits high digestibility, making nutrients readily available for absorption and utilization.

Fish constitute the primary dietary source of **omega-3 long-chain polyunsaturated fatty acids**, particularly eicosapentaenoic acid and docosahexaenoic acid. These fatty acids provide crucial benefits for cardiovascular health, brain development and function, visual acuity, and immune response. Human bodies synthesize these compounds inefficiently from plant-derived precursors, making dietary intake essential for optimal health. Regular fish consumption associates with reduced cardiovascular disease risk, improved cognitive function, better pregnancy outcomes, and other health benefits. Pisciculture that produces omega-3 rich fish provides these nutritional benefits at scale.

Fish supply essential micronutrients including vitamin A, vitamin D, vitamin B12, iron, zinc, calcium, iodine, and selenium. These micronutrients are critical for growth, development, immune function, and prevention of deficiency diseases. **Hidden hunger**—micronutrient deficiencies despite adequate caloric intake—affects billions of people worldwide, particularly women and children in developing nations. Small fish consumed whole with bones provide especially rich micronutrient sources. Integrating fish consumption into diets can substantially reduce micronutrient deficiencies, improving health outcomes especially for vulnerable populations.

The bioavailability of nutrients from fish generally exceeds that from plant sources. Iron from fish, for example, is absorbed more efficiently than iron from plant foods. The protein quality of fish surpasses most plant proteins. These bioavailability advantages mean fish can address nutritional deficiencies more effectively per unit consumed compared to many alternatives.

Economic Contributions and Livelihood Opportunities

Pisciculture generates substantial economic value through direct production, employment creation, income generation, foreign exchange earnings, and support for ancillary industries. The economic importance extends from subsistence farming providing household food security and modest cash income to industrial-scale operations generating hundreds of millions in annual revenue.

Direct production value of global aquaculture exceeded 250 billion US dollars in recent years, representing substantial economic activity.

This value reflects only farm-gate prices; further value addition occurs through processing, distribution, and retail marketing. The economic multiplier effects of aquaculture generate substantially more total economic activity than direct production value alone.

Pisciculture provides **livelihood opportunities** for millions of people globally, including smallholder farmers, industrial workers, input suppliers, processors, traders, and service providers. In many developing nations, aquaculture represents an important pathway out of poverty for rural households with limited land or employment alternatives. Small-scale fish farming can be integrated with agriculture, utilizing family labor during slack agricultural seasons and generating supplementary income. The relatively low capital requirements for basic pond aquaculture enable resource-poor farmers to participate.

The aquaculture sector creates diverse **employment opportunities** requiring various skill levels from manual labor to advanced scientific and technical expertise. Production operations employ pond attendants, feeding staff, harvest workers, and farm managers. Hatcheries need skilled technicians capable of managing delicate breeding and larval rearing processes. Feed mills employ nutritionists, mill operators, and quality control staff. Equipment manufacturing, engineering services, veterinary services, financial services, research institutions, and government agencies all provide aquaculture-related employment.

Export earnings from aquaculture products contribute significantly to national economies, particularly in developing nations where foreign exchange may be scarce. Shrimp farming countries earn billions annually from exports to developed nation markets. Salmon farming generates major export revenues for Norway, Chile, Scotland, and Canada. Tilapia, catfish, and other species increasingly enter international trade. These export revenues support economic development, technology acquisition, and debt servicing.

Development of aquaculture value chains creates business opportunities beyond farm production. **Feed manufacturing** has grown into a substantial industry supplying nutritionally complete diets to intensive aquaculture operations. Specialized equipment suppliers provide aeration devices, pumps, nets, cages, water quality testing instruments, and other necessary inputs. Processing facilities add value through cleaning, filleting, freezing, smoking, and packaging operations. Transportation, cold storage, and distribution services link producers with consumers. Financial institutions provide credit for farm establishment and operation. This complex value chain generates economic activity and employment at multiple levels.

Technology development and innovation associated with aquaculture create additional economic value. Research institutions develop improved culture practices, genetic strains, health management protocols, and production systems. Private companies commercialize these innovations, generating revenue while advancing industry capabilities. Intellectual property associated with genetic improvements, patented technologies, and proprietary knowledge represents economic value created by aquaculture sector innovation.

Economic Benefits at Multiple Scales

The economic impacts of pisciculture manifest at household, community, regional, and national scales. At household level, fish farming provides food security, nutritional diversity, and cash income. Even modest production can substantially improve household welfare. The flexibility of aquaculture systems enables integration with other activities, optimizing use of family labor and resources.

At community level, aquaculture development stimulates local economic activity. Establishment of multiple fish farms creates demand for inputs purchased from local suppliers, generates employment for community members, and may support development of local processing and marketing infrastructure. These multiplier effects amplify direct economic impacts.

Regional economies benefit when aquaculture development achieves sufficient scale to support specialized service provision. Feed mills, hatcheries, equipment suppliers, and processing facilities serve multiple farms across a region, creating employment and economic activity. Successful aquaculture development can transform regional economies, shifting from subsistence agriculture toward more dynamic market-oriented production systems.

National economic impacts include production value contribution to gross domestic product, generation of tax revenues, foreign exchange earnings from exports, and reduced foreign exchange expenditure on imported fish products. Countries achieving aquaculture success experience measurable economic benefits evident in national statistics.

Social and Gender Dimensions

Pisciculture development generates social benefits extending beyond purely economic considerations. In many societies, fish farming empowers marginalized groups including women, youth, and ethnic minorities by providing accessible income opportunities. Women often participate extensively in aquaculture activities including feeding, monitoring, processing, and marketing. Recognition of women's contributions and ensuring equitable benefit distribution remain important development considerations.

Youth employment in aquaculture addresses rural unemployment and underemployment affecting many developing regions. Modern aquaculture can be attractive to young people if presented as a technologically sophisticated, innovative sector rather than traditional subsistence farming. Value chain development creates diverse employment opportunities suitable for various educational backgrounds and interests.

Aquaculture development can promote social cohesion through farmer organizations, cooperatives, and producer groups. These collective institutions facilitate technology transfer, input acquisition, marketing, and advocacy. Successful cooperatives often provide social benefits including mutual support, knowledge sharing, and community identity.

1.4 Types of Fish Culture – Extensive, Semi-Intensive, and Intensive Systems

Fish culture systems exhibit enormous diversity in operational characteristics, resource requirements, management complexity, and production outputs. This diversity reflects adaptation to varying environmental conditions, available resources, technical capabilities, market demands, and cultural contexts. Classification of culture systems along an intensity gradient from extensive to intensive provides a useful framework for understanding this diversity and evaluating system suitability for particular circumstances.

The intensity of aquaculture systems fundamentally relates to the degree of human intervention, input application, environmental control, and management attention devoted to fish production. This intensity continuum spans from minimal intervention systems largely dependent on natural productivity to highly controlled intensive systems where virtually all environmental factors and inputs are managed deliberately. Understanding characteristics, advantages, limitations, and appropriate applications of systems along this intensity gradient enables informed selection of culture approaches suited to specific situations.

Extensive Fish Culture Systems

Extensive fish culture represents the lowest intensity approach, characterized by minimal human intervention, reliance on natural productivity, low stocking densities, minimal external inputs, and modest production outputs per unit area. These systems typically involve large water bodies including natural or artificial ponds, lakes, reservoirs, or seasonal floodplain waters where fish populations are established and subsequently harvested after a growing period. Management activities are limited, often restricted to initial stocking, occasional monitoring, and final harvest.

The fundamental principle underlying extensive systems involves exploiting natural food production within water bodies to support fish growth. Phytoplankton, zooplankton, benthic organisms, aquatic insects, and detritus produced through natural ecological processes provide nutrition for cultured fish. Stocking densities must remain low enough that natural food production satisfies fish nutritional requirements without supplementation. When stocking densities exceed the carrying capacity supported by natural productivity, growth rates decline and fish may fail to reach marketable size within acceptable timeframes.

Extensive systems demonstrate several notable advantages. **Capital requirements** are minimal since expensive infrastructure, equipment, and inputs are largely unnecessary. Simple earthen ponds or utilization of existing water bodies eliminates major construction costs. The absence of feeding, aeration, and intensive management reduces operational costs substantially. These economic characteristics make extensive aquaculture accessible to resource-poor farmers unable to invest in more intensive approaches.

Technical knowledge requirements for extensive systems are relatively modest. Farmers need basic understanding of species biology, stocking procedures, and harvest methods but need not master complex nutrition, water quality management, or health protocols required in intensive systems. This simplicity facilitates adoption by farmers with limited formal education or technical training.

Environmental sustainability of extensive systems generally exceeds more intensive alternatives. The reliance on natural productivity minimizes external input requirements and associated environmental impacts. Waste discharge remains low since fish production depends on assimilation of existing nutrients rather than supplementary feeding. Integration with natural ecosystems typically generates fewer conflicts than intensive systems requiring substantial infrastructure modification.

However, extensive systems present significant limitations constraining their applicability and productivity. **Production output per unit area** is inherently low, typically ranging from a few hundred to perhaps one thousand kilograms per hectare annually. These modest yields reflect the limited natural productivity of most aquatic ecosystems. Increasing production requires expanding culture area rather than intensifying production on existing land, which may not be feasible where land and water resources are scarce.

Production predictability and control are limited in extensive systems. Natural productivity fluctuates seasonally and annually depending on climatic conditions, nutrient availability, and ecological dynamics largely beyond farmer control. This variability creates production uncertainty affecting planning and market supply. Diseases, predators, and competitors impact fish populations without effective management intervention possible in extensive systems.

Time to harvest is typically prolonged in extensive systems due to slow growth rates on natural diets. Fish may require two or more years to reach market size, extending production cycles and delaying returns on investment. Long production cycles increase risks from natural disasters, disease outbreaks, predation, and market price fluctuations.

Species suitable for extensive culture generally exhibit tolerance for variable environmental conditions, efficient utilization of natural foods, and adequate growth on low-quality diets. **Common carp** represents a classical extensive culture species, thriving in diverse environments and feeding omnivorously on natural organisms. **Tilapia** species successfully utilize phytoplankton-based natural productivity. Various indigenous species adapted to local conditions often perform well in extensive systems.

Extensive culture remains important in many contexts despite productivity limitations. In regions with abundant land and water resources but limited capital and technical capacity, extensive systems provide accessible entry points into aquaculture. Rural areas where fish farming serves primarily for household consumption rather than commercial sale may appropriately employ extensive approaches. **Culture-based fisheries** in large reservoirs, lakes, and seasonal water bodies often utilize extensive approaches, enhancing fisheries productivity while maintaining essentially natural ecosystem characteristics.

Semi-Intensive Fish Culture Systems

Semi-intensive culture represents an intermediate intensity level incorporating moderate management intervention, supplementary feeding, modest external inputs, medium stocking densities, and substantially higher production outputs compared to extensive systems while avoiding the high costs and complexity of fully intensive approaches. These systems blend natural productivity with managed inputs, achieving production enhancement while maintaining economic accessibility and environmental sustainability.

The defining characteristic of semi-intensive systems is **supplementary feeding** providing additional nutrition beyond natural productivity. Fish receive formulated feeds or agricultural byproducts at rates satisfying some but not all nutritional requirements, with natural foods continuing to contribute significantly to growth. Supplementary feeding enables increased stocking densities and accelerated growth while maintaining lower feeding costs compared to complete reliance on formulated diets. Typical supplementary feeds include rice bran, wheat bran, oil cakes, brewery wastes, and other agricultural processing byproducts available at relatively low cost.

Stocking densities in semi-intensive systems typically range from moderate levels perhaps five to ten times higher than extensive systems but substantially below intensive culture densities. These intermediate densities balance production potential against natural productivity, feed costs, and management capacity. Polyculture systems combining multiple compatible species often characterize semi-intensive operations, efficiently utilizing available food resources occupying different ecological niches within the culture environment.

Fertilization with organic manures or inorganic fertilizers commonly supplements natural productivity in semi-intensive systems. Fertilizers increase primary productivity by alleviating nutrient limitations on phytoplankton and plant growth, thereby enhancing food availability for fish. Organic fertilizers including animal manures, compost, and green manures provide dual benefits: supplying nutrients for phytoplankton production while also contributing organic matter directly consumed by some fish species. Inorganic fertilizers provide specific nutrients in readily available forms, enabling targeted correction of limiting nutrients.

Water management in semi-intensive systems receives more attention than extensive operations but remains simpler than intensive systems. Periodic water exchange maintains acceptable water quality by diluting accumulated metabolites and replenishing dissolved oxygen. Lime application may correct pH and alkalinity, improving water chemistry for fish and natural productivity. These management interventions are intermittent rather than continuous, requiring modest equipment and technical knowledge.

Infrastructure requirements for semi-intensive systems remain moderate. Earthen ponds continue dominating, though construction standards may be higher than extensive operations to facilitate better water control. Basic infrastructure including inlet and outlet structures, drainage systems, and harvest facilities are necessary. Simple equipment like nets, buckets, and weighing scales suffice for routine operations. Some farms may employ mechanical feeders or basic aeration devices but extensive mechanization is not typical.

Production outputs from semi-intensive systems typically range from two to five thousand kilograms per hectare annually, representing substantial enhancement over extensive systems while remaining well below intensive system yields. These intermediate production levels generate meaningful income for farm families while maintaining economic viability without extreme capital investment or technical sophistication.

The **economic profile** of semi-intensive systems balances input costs against production revenues. Supplementary feeding, fertilization, and enhanced management increase operational costs compared to extensive systems. However, substantially higher production outputs typically generate favorable returns on these incremental investments. The economic risk remains moderate since capital investment is not excessive and some production occurs even if management is imperfect. This risk profile suits small-scale commercial farmers seeking income generation while avoiding excessive financial exposure.

Environmental implications of semi-intensive systems fall between extensive and intensive extremes. Fertilization and feeding add nutrients to culture environments, with some proportion discharged in effluent. However, nutrient loading remains moderate compared to intensive systems. Integration with agriculture through use of animal manures and crop byproducts creates beneficial nutrient cycling, converting agricultural wastes into fish protein. When properly managed, semi-intensive systems maintain reasonable environmental sustainability.

Technical knowledge requirements for semi-intensive culture exceed extensive systems but remain accessible to farmers with basic education and appropriate training. Understanding species biology, feeding practices, fertilization regimes, water quality principles, and health management basics enables competent management. Extension services, farmer training programs, and peer learning support technology adoption.

Semi-intensive aquaculture dominates in many developing Asian nations where it effectively balances productivity, accessibility, and sustainability. Small-scale farmers operating family ponds typically employ semi-intensive approaches, producing fish for both household consumption and market sale. The economic returns support rural livelihoods while resource requirements remain within reach of resource-constrained farm households. This accessibility and productivity combination explains the prevalence and continued importance of semi-intensive systems in global aquaculture.

Intensive Fish Culture Systems

Intensive fish culture represents the highest level of management intervention, input application, environmental control, and production optimization. These systems employ high stocking densities, complete nutritionally balanced feeds, sophisticated water quality management, disease prevention protocols, and often advanced environmental control technologies. Production outputs substantially exceed extensive and semi-intensive systems, but capital costs, technical requirements, and operational complexity are correspondingly higher.

The fundamental principle of intensive culture involves creating and maintaining optimal conditions for maximum fish growth and survival through comprehensive environmental management. Natural productivity contributes negligibly to fish nutrition; complete feeds providing all nutritional requirements are supplied in amounts closely matched to fish appetite and metabolic capacity. High stocking densities are sustained through vigorous water quality management preventing accumulation of toxic metabolites and maintaining adequate dissolved oxygen. Diseases and parasites are controlled through biosecurity, health monitoring, and preventive treatments.

Stocking densities in intensive systems may exceed 50 to 100 kilograms per cubic meter of culture volume, multiple orders of magnitude higher than extensive systems. Such densities are sustainable only through continuous management intervention preventing water quality deterioration that would otherwise cause mass mortality. The economic logic of intensive systems depends on achieving high productivity per unit of expensive infrastructure, justifying the substantial investments and operational costs.

Complete formulated feeds constitute a major operational cost in intensive systems, typically representing 40 to 60 percent of total production expenses. These feeds are nutritionally complete, carefully formulated to provide optimal ratios of protein, carbohydrates, lipids, vitamins, minerals, and other essential nutrients. Feed composition is tailored to species requirements and life stages, with different formulations for larvae, juveniles, and grow-out. High-quality feeds support rapid growth and efficient feed conversion, minimizing production cycle duration and maximizing economic returns.

Water quality management in intensive systems requires continuous attention and often sophisticated technology. Dissolved oxygen must be maintained near saturation through vigorous aeration or pure oxygen injection. Ammonia, nitrite, carbon dioxide, and other potentially toxic metabolites are controlled through water exchange, biofiltration, or chemical treatment. Temperature, pH, salinity, and other parameters are monitored and adjusted to maintain optimal ranges. Automated sensors, controllers, and alarms enable continuous monitoring and rapid response to problems.

Several intensive culture system types exemplify different approaches to environmental management. **Flow-through systems** including raceways and tanks achieve water quality maintenance through continuous inflow of clean water and discharge of used water. High water flow rates dilute metabolites and replenish oxygen, supporting high fish densities. These systems require abundant supplies of high-quality source water, limiting their geographical applicability but offering simplicity and reliability where water is available.

Recirculating aquaculture systems employ sophisticated water treatment technologies enabling reuse of culture water with minimal makeup water requirements. Mechanical filtration removes solid wastes, biological filtration oxidizes ammonia to less toxic nitrate, and other treatment processes maintain water quality. Recirculating systems can be established virtually anywhere regardless of natural water availability, enabling urban aquaculture, desert production, or facilities in cold climates with controlled environments. However, these systems require substantial capital investment in treatment equipment and technical expertise for successful operation.

Cage culture in lakes, reservoirs, or marine environments represents another intensive approach, confining high-density fish populations in floating or submerged enclosures while relying on natural water bodies for waste assimilation and oxygen replenishment. Current flow through cages removes metabolic wastes and supplies oxygen, enabling stocking densities comparable to land-based intensive systems without water treatment infrastructure. Cage culture dominates salmon farming and increasingly is employed for other species in suitable water bodies.

Production outputs from intensive systems commonly exceed 50 to 150 kilograms per cubic meter for finfish species, with some systems achieving even higher densities for particularly tolerant species. When expressed per unit land area, yields may reach tens of thousands of kilograms per hectare annually. These dramatic productivity levels enable economic viability despite high operational costs and justify the substantial infrastructure investments.

The **economic profile** of intensive systems involves high capital requirements for infrastructure, equipment, and initial stocking, along with substantial ongoing operational costs for feed, energy, labor, and maintenance. However, high production outputs generate significant revenues. When well managed, intensive systems yield attractive profit margins and returns on investment. The economic viability is sensitive to production efficiency, market prices, and cost control. Technical problems causing mortality, slow growth, or disease can rapidly erode profitability, creating financial risk.

Technical requirements for intensive culture are substantial. Managers need comprehensive understanding of fish biology, nutrition, water chemistry, engineering systems, health management, and business operations. Trained technical staff, often including specialists in different domains, are necessary for larger operations. Access to diagnostic services, veterinary support, and equipment maintenance is essential. The technical complexity limits participation to well-capitalized investors capable of employing qualified personnel and implementing sophisticated management systems.

Environmental considerations surrounding intensive aquaculture generate ongoing debate. Concentrated wastes from high-density culture create potential pollution if not properly managed. Feed manufacturing requires substantial quantities of protein sources, often from wild fish, raising sustainability concerns. Disease outbreaks in intensive systems may require chemical treatments with environmental implications. However, modern intensive systems increasingly employ best management practices, waste treatment, integrated production, and environmental monitoring to minimize negative impacts. The land and water efficiency of intensive systems means they can produce large quantities of fish with relatively small environmental footprint per unit produced, particularly compared to extensive systems requiring large land areas.

Intensive culture dominates production of salmonids, marine species, and increasingly other species where market prices justify the costs and complexity. Industrial-scale operations producing thousands of tons annually supply global markets with consistent quality products. The technological sophistication and productivity of intensive systems will likely continue increasing as innovation advances and environmental regulations tighten.

Comparative Analysis and System Selection

Selection among extensive, semi-intensive, and intensive culture approaches depends on numerous factors including available resources, technical capacity, target species, market characteristics, environmental conditions, and production objectives. No single approach is universally superior; each offers advantages in appropriate contexts.

Extensive systems suit situations with abundant land and water but limited capital and technical resources. They provide accessible entry into aquaculture for resource-poor farmers and enable sustainable enhancement of natural water bodies. However, low productivity limits income generation potential.

Semi-intensive systems offer favorable balance of accessibility, productivity, and sustainability for small-scale commercial production. They generate meaningful income while remaining within technical and financial reach of small farmers. Semi-intensive approaches dominate in many developing nations for these reasons.

Intensive systems maximize productivity from limited resources, enabling large-scale commercial production supplying substantial market demand. They require significant capital, technical expertise, and sophisticated management but offer potential for high returns. Industrial aquaculture increasingly employs intensive approaches to achieve the efficiency and scale demanded by global markets.

Many successful aquaculture operations combine elements from different intensity levels, adapting systems to specific circumstances. Understanding the characteristics, advantages, and limitations of systems along the intensity continuum enables informed decisions optimizing production outcomes given available resources and constraints.

Table 1: Comparative Characteristics of Fish Culture Systems

Parameter	Extensive	Semi-Intensive	Intensive

Stocking Density	Very Low (< 5,000/ha)	Moderate (10,000–50,000/ha)	Very High (> 100,000/ha)
Feeding Practice	Natural food only	Supplementary feeding	Complete formulated feed
Water Management	Minimal exchange	Periodic exchange	Continuous flow/ treatment
Production Output	500–1,500 kg/ ha/yr	2,000–5,000 kg/ ha/yr	> 50,000 kg/ha/yr
Capital Requirement	Very Low	Moderate	Very High
Technical Skill	Basic	Intermediate	Advanced
Infrastructure	Simple ponds	Improved ponds	Tanks, cages, RAS
Environmental Control	Negligible	Partial	Comprehensive

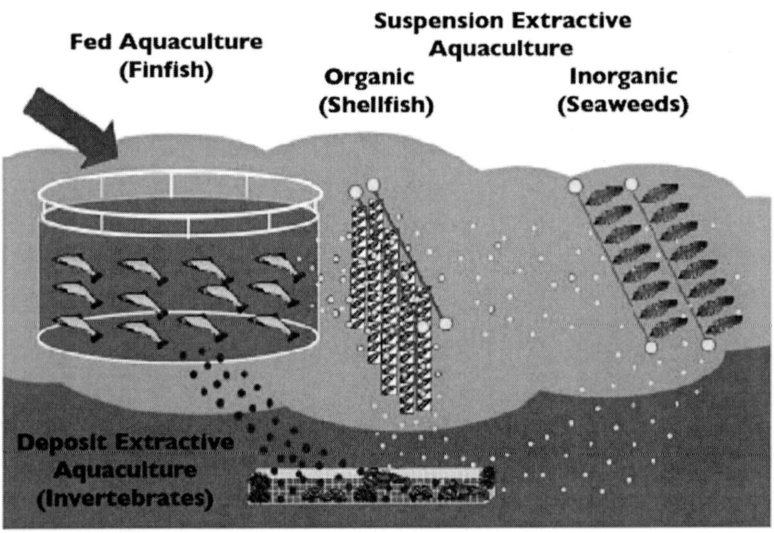

Figure. 1: Extensive Culture System Configuration

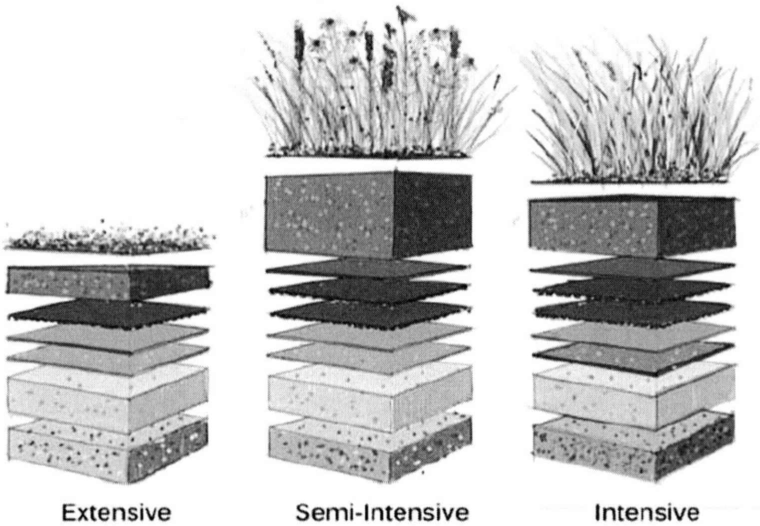

Figure. 2: Semi-Intensive Polyculture System

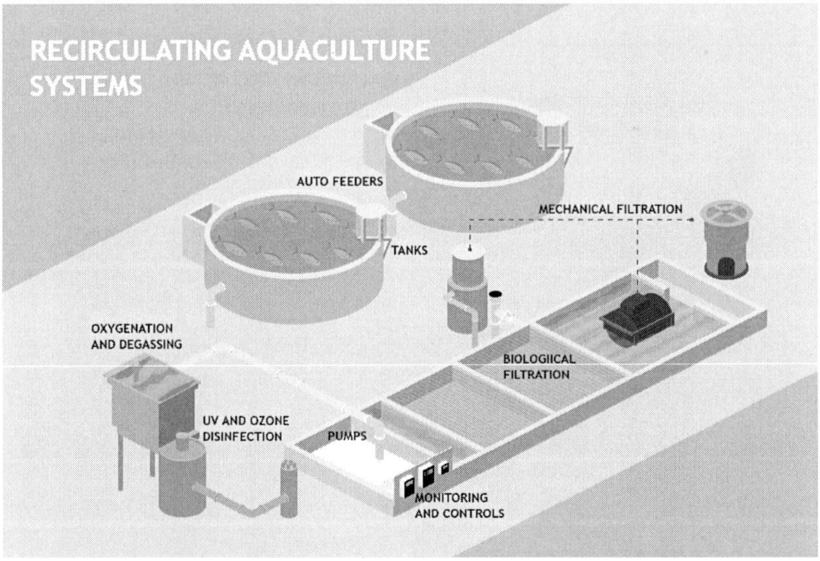

Figure. 3: Intensive Recirculating Aquaculture System

1.5 Principles of Applied Fish Farming

Successful applied fish farming rests on fundamental principles integrating biological understanding with practical management, economic viability, and environmental sustainability. These principles transcend specific culture systems, species, or geographical contexts, providing universal guidance for designing, implementing, and operating fish production operations. Mastery of these principles enables practitioners to make informed decisions, troubleshoot problems, and continuously improve performance.

Biological Principles

The foundation of successful fish culture lies in understanding and accommodating the biological requirements and characteristics of cultured species. Fish are living organisms whose survival, growth, reproduction, and health depend on maintaining conditions within tolerable ranges and optimally within preferred ranges for their specific biology.

Species-Environment Compatibility represents the most fundamental principle. Each fish species evolved under particular environmental conditions and possesses specific tolerances and preferences for temperature, dissolved oxygen, pH, salinity, water chemistry, and other factors. Successful culture requires selecting species suited to available environmental conditions or modifying environments to meet species requirements. Attempting to culture species in unsuitable environments typically results in poor growth, high mortality, or production failure regardless of other management efforts.

Temperature profoundly influences fish metabolism, growth, feed consumption, oxygen requirements, disease susceptibility, and reproductive function. **Temperature-Growth Relationships** follow predictable patterns within species-specific tolerance ranges. Below optimal temperatures, metabolism slows and growth rates decline. Above optimal temperatures, stress increases, oxygen requirements may exceed availability, and mortality risks rise. Understanding these relationships enables appropriate species selection, production timing, and management adjustments responding to temperature variation.

Oxygen Requirements are critical for fish survival and optimal growth. Fish extract dissolved oxygen from water through their gills, with oxygen consumption rates depending on species, size, temperature, feeding level, and activity. Warm-water species generally tolerate lower dissolved oxygen levels than cold-water species, but all species exhibit reduced growth and increased stress at suboptimal oxygen levels. Understanding oxygen dynamics including sources through photosynthesis and atmospheric diffusion, sinks through respiration and decomposition, and management through aeration enables maintenance of adequate oxygen levels supporting high productivity.

Feeding Biology and Nutrition fundamentally govern fish growth and production efficiency. Fish require balanced nutrition providing energy, protein, essential fatty acids, vitamins, minerals, and other nutrients in appropriate ratios. Natural feeding habits influence culture management; herbivorous species may utilize phytoplankton or supplementary plant materials, carnivorous species require high-protein diets, and omnivorous species accept diverse feed types. Understanding nutritional requirements, digestive physiology, feeding behavior, and feed conversion efficiency enables formulation of cost-effective feeding strategies maximizing growth while minimizing waste.

Growth Patterns and Production Cycles follow predictable trajectories influenced by genetics, nutrition, environmental conditions, and stocking density. Growth is typically rapid during early life stages and declines as fish approach maturity. Understanding growth dynamics enables optimizing harvest timing when growth rates decline to the point where continued feeding becomes economically inefficient. Production planning based on realistic growth expectations prevents unrealistic goals and disappointment.

Reproductive Biology understanding is essential for successful seed production or breeding programs. Fish reproduction varies enormously among species, from external fertilization in open water to elaborate nest-building and parental care. Many species require specific environmental cues triggering spawning including temperature changes, photoperiod, lunar cycles, or water flow. Successful hatchery operations accommodate these biological requirements through environmental manipulation or hormone-induced breeding techniques.

Population Dynamics and Density Effects influence culture outcomes through competition, behavioral interactions, and disease transmission. Overstocking beyond carrying capacity causes growth stunting through competition for food and space, increased stress, deteriorating water quality, and elevated disease risk. Understanding appropriate stocking densities for particular systems and management levels prevents these problems while maximizing production.

Stress Physiology profoundly affects fish health, growth, and survival. Stress from poor water quality, handling, crowding, transport, or other factors triggers physiological responses that suppress immune function, reduce growth, and increase disease susceptibility. Minimizing stress through proper handling, avoiding sudden environmental changes, and maintaining optimal conditions enhances production success.

Environmental Management Principles

Environmental conditions within culture systems must be maintained within ranges supporting fish health and optimal productivity. Environmental management principles provide guidance for creating and maintaining suitable conditions.

Water Quality Management encompasses maintenance of dissolved oxygen, control of toxic metabolites, regulation of pH and alkalinity, and management of suspended solids. The principle of **Oxygen Balance** requires that oxygen inputs through photosynthesis, aeration, or water exchange must equal or exceed oxygen consumption by fish, decomposition, and chemical oxidation. Violation of this balance causes oxygen depletion and potential mortality. Similarly, **Waste Assimilation Capacity** principles recognize that culture environments can process and assimilate certain waste loads through biological processes, but excessive loading overwhelms assimilative capacity causing toxic accumulation.

Carrying Capacity concepts are fundamental for sustainable culture. Every culture system has a maximum biomass supportable given its physical characteristics, management intensity, and environmental conditions. Stocking beyond carrying capacity leads to degraded conditions and poor production outcomes. Proper management maintains stocking levels within carrying capacity, adjusting as conditions change seasonally or as fish grow.

Water Management Strategies must balance competing objectives including maintaining quality, conserving resources, managing costs, and preventing disease introduction. Water exchange dilutes metabolites and replenishes oxygen but requires source water availability, energy for pumping, and care to prevent pathogen introduction. Aeration supplies oxygen without water exchange but requires energy and equipment investment. Biological filtration processes wastes but requires space and management. Optimizing water management involves selecting appropriate combinations of strategies suited to system characteristics and constraints.

Nutrient Cycling and Productivity management applies particularly to semi-intensive systems relying partially on natural productivity. The principle recognizes that nutrients enter ponds through feeds and fertilizers, are transformed through ecological processes, and ultimately must be removed through harvest or discharge to prevent excessive accumulation. Balanced fertilization and feeding maintain productive systems without water quality deterioration.

Environmental Monitoring provides essential feedback enabling responsive management. Regular measurement of temperature, dissolved oxygen, pH, ammonia, nitrite, alkalinity, and other parameters detects problems before fish health seriously deteriorates. Monitoring frequency and parameters measured should match system intensity and risk levels.

Production Management Principles

Effective production management integrates biological understanding with operational planning, resource allocation, and performance monitoring to achieve production goals efficiently.

Stocking Management begins with selection of appropriate species suited to environment, market demand, and management capabilities. Quality seed from reputable sources ensures good growth potential, disease freedom, and genetic suitability. Proper stocking density balances production potential against carrying capacity and management level. Timing of stocking influences production cycles, market timing, and environmental conditions during critical early growth periods.

Feeding Management profoundly affects production economics since feed typically constitutes the largest operational cost. The principle of **Feed Conversion Optimization** guides feeding practices toward maximizing growth per unit feed consumed. Underfeeding limits growth below potential; overfeeding wastes expensive feed and degrades water quality. Optimal feeding matches feed quantity and quality to fish appetite and nutritional requirements, adjusting for fish size, temperature, and condition. Regular monitoring of feeding response and growth rates enables adaptive management.

Growth Performance Monitoring through periodic sampling, weighing, and measuring provides essential feedback on production progress. Comparing actual versus expected growth identifies problems or opportunities for management improvement. Monitoring also guides decisions regarding feeding adjustments, harvest timing, and production forecasting.

Health Management emphasizes disease prevention through biosecurity, stress minimization, and environmental optimization rather than reactive disease treatment. Biosecurity protocols prevent pathogen introduction through contaminated water, infected seed stock, equipment, or personnel. Maintaining optimal environmental conditions and minimizing stress support robust immune function reducing disease susceptibility. Early detection through behavioral observation and health monitoring enables prompt response when problems occur.

Harvest Management timing influences product quality, market price, and economic returns. Harvesting at optimal size for target markets maximizes value. Proper harvest techniques minimize stress and physical damage maintaining product quality. Efficient harvest methods reduce labor costs and time requirements. Post-harvest handling including icing, processing, and transport significantly affect final product quality and value realization.

Economic and Business Principles

Fish farming is fundamentally an economic activity seeking profitable returns on invested resources. Economic principles guide resource allocation, investment decisions, and operational management toward financial sustainability and competitiveness.

Production Economics principles recognize that profit results from the difference between revenues and costs. Maximizing profit requires optimizing this difference rather than simply maximizing production volume. Understanding production functions relating inputs to outputs enables identifying optimal input levels where marginal returns equal marginal costs. Beyond this point, additional inputs reduce profit despite increasing output.

Cost Management involves controlling both fixed costs for infrastructure and equipment and variable costs including seed, feed, labor, and utilities. Efficient operations minimize costs per unit produced through economies of scale, optimal input use, and productivity improvement. Regular cost tracking and analysis identifies cost-saving opportunities and detects problems early.

Market Orientation requires aligning production with market demand rather than simply producing whatever is technically feasible. Understanding consumer preferences, seasonal demand patterns, price dynamics, and market channels enables strategic decisions regarding species selection, production timing, product forms, and marketing approaches. Value addition through processing, branding, or direct marketing may significantly enhance returns.

Risk Management addresses the numerous uncertainties inherent in aquaculture including biological risks from disease, predation, or poor growth; environmental risks from weather extremes, water supply disruptions, or pollution; market risks from price fluctuations or demand changes; and operational risks from equipment failures or labor problems. Risk mitigation strategies include diversification, insurance, contingency planning, conservative financial management, and technical excellence reducing biological and environmental risks.

Financial Planning and Management ensures adequate capital for investment and operations, manages cash flow through production cycles, and evaluates financial performance. Proper budgeting, record-keeping, and financial analysis support informed decision-making and enable demonstrating creditworthiness for financing.

Sustainability Principles

Sustainable aquaculture meets present production and economic goals without compromising ability of future generations to meet their needs. Sustainability principles integrate environmental, social, and economic considerations.

Environmental Stewardship requires minimizing negative environmental impacts including water pollution, habitat destruction, and resource depletion. Best management practices reduce waste discharge, protect water quality, and prevent ecosystem degradation. Resource efficiency in water use, energy consumption, and feed utilization reduces environmental footprint. Environmental monitoring and adaptive management ensure ongoing environmental compatibility.

Social Responsibility encompasses fair treatment of workers, contribution to community welfare, and respect for local culture and traditions. Equitable benefit distribution ensures that aquaculture development improves livelihoods broadly rather than concentrating benefits narrowly. Engagement with communities and responsiveness to concerns builds social acceptance and long-term viability.

Resource Use Efficiency recognizes that inputs including land, water, feed, energy, and capital are finite resources that should be used efficiently. Maximizing production per unit resource input reduces both economic costs and environmental impacts. Continuous improvement in resource efficiency through technology adoption and management refinement advances sustainability.

Long-term Perspective in planning and management prevents short-term decisions that compromise long-term sustainability. Maintaining soil and water quality, preserving genetic resources, avoiding overexploitation of natural seed sources, and building operational capacity support continued productivity over decades rather than seeking maximum short-term extraction followed by collapse.

These fundamental principles provide a framework for successful applied fish farming. Mastery and application of these principles enable practitioners to design appropriate systems, implement effective management, solve problems, and continuously improve performance toward productive, profitable, and sustainable operations.

1.6 Global and Regional Trends in Pisciculture Development

Contemporary pisciculture exists within a dynamic context of technological innovation, evolving market demands, environmental constraints, and socioeconomic changes. Understanding current trends and future trajectories is essential for practitioners, investors, policymakers, and students seeking to participate effectively in this rapidly evolving sector. This section examines major global trends and regional patterns shaping pisciculture development.

Global Production Trends and Projections

Global aquaculture production has expanded dramatically over recent decades, growing at approximately 5 to 6 percent annually—substantially faster than terrestrial animal agriculture or capture fisheries. This growth trajectory has transformed aquaculture from a marginal activity to a major global food production system. Recent global production exceeds 110 million tons annually including finfish, crustaceans, mollusks, and aquatic plants, with finfish representing the largest component.

Production growth has occurred unevenly across regions and species. **Asian dominance** is overwhelming, with the region producing over 90 percent of global aquaculture output. China alone accounts for approximately 60 percent of global production, reflecting decades of government support, favorable natural conditions, large rural population providing labor, and cultural acceptance of aquaculture products. Other Asian producers including India, Indonesia, Vietnam, Bangladesh, and Thailand contribute substantially to regional and global totals.

Production outside Asia remains comparatively modest but is growing in certain regions and species. **Latin American aquaculture**, particularly Chilean salmon farming and Ecuadorian shrimp production, has expanded significantly. **African aquaculture** is growing rapidly from a small base, driven by population growth, protein demand, and development initiatives, though production remains a tiny fraction of Asian output. **European aquaculture** focuses primarily on high-value species including salmonids and marine fishes, with production constrained by environmental regulations, high costs, and limited suitable sites.

Species diversification characterizes recent development. While traditional species including carps, tilapias, catfishes, and salmonids continue dominating production volumes, successful commercialization of additional species broadens industry portfolio. Marine species culture is expanding, reducing pressure on wild stocks and supplying growing demand. Innovative species selection seeking fish suited to particular environmental niches, market segments, or production systems creates opportunities while reducing dependence on a few dominant species.

Intensification and Technological Innovation

Aquaculture production has intensified significantly, with increasing proportions produced in semi-intensive and intensive systems employing advanced technologies and management practices. This intensification trend reflects multiple drivers including land and water scarcity, need for higher productivity, market demand for consistent quality, and environmental regulations limiting extensive system expansion.

Recirculating aquaculture systems represent perhaps the most dramatic intensification trend. These highly controlled systems enable fish production virtually anywhere regardless of natural water availability, climate, or traditional geographic limitations. While capital costs are substantial and technical requirements demanding, successful recirculating systems demonstrate remarkable productivity, biosecurity, and environmental performance. The technology continues advancing with improved biofilter designs, energy efficiency, waste recovery systems, and automation. Recirculating system adoption is growing particularly in developed nations with high land costs, environmental standards, and market premiums for locally produced or sustainably grown products.

Precision aquaculture employs digital technologies, sensors, data analytics, and automation to optimize production processes. Automated feeding systems adjust feed delivery based on real-time fish behavior and appetite, improving feed conversion while reducing waste. Sensor networks continuously monitor water quality parameters, alerting managers to problems and enabling automated corrective responses. Underwater cameras observe fish health, growth, and behavior, detecting problems early. Data analytics integrate information from multiple sources, identifying patterns and optimizing management decisions. Artificial intelligence applications are emerging for disease diagnosis, growth prediction, and operational planning. These technologies promise to enhance productivity, reduce environmental impacts, and improve economic returns while potentially reducing labor requirements.

Genetic improvement through selective breeding has accelerated dramatically with application of modern genetic technologies. Selective breeding programs now produce genetically improved strains of major species exhibiting substantially superior growth rates, feed conversion efficiency, disease resistance, or other economically important traits compared to wild populations or earlier generations. The **GIFT tilapia** strain, developed through international collaboration, exemplifies the transformative impact of systematic selective breeding, demonstrating 60 to 80 percent growth advantage over wild tilapia after multiple generations of selection. Similar improvement programs target carps, salmonids, shrimps, and other major species.

Molecular genetics tools accelerate genetic improvement through marker-assisted selection, genomic selection, and eventually gene editing, though regulatory and social acceptance issues surround advanced biotechnologies. Understanding of fish genomes enables identifying genes controlling important traits, selecting breeding candidates based on genetic markers rather than only observed phenotypes, and potentially modifying genomes directly to introduce desired characteristics.

Sustainability Emphasis and Environmental Standards

Sustainability considerations increasingly influence aquaculture development as environmental awareness grows, regulations tighten, and market demand for responsibly produced products increases. The industry faces legitimate criticisms regarding pollution, habitat destruction, disease transmission to wild populations, and resource use, motivating efforts toward more sustainable practices.

Waste management technologies and practices are advancing to reduce pollution from intensive aquaculture. Solid waste collection systems remove feces and uneaten feed before decomposition degrades water quality. Biological treatment systems process collected wastes, potentially recovering nutrients for agricultural use or producing biogas for energy. Constructed wetlands treat aquaculture effluent before discharge, removing nutrients and improving water quality. Zero-discharge systems incorporating treatment technologies eliminate direct effluent discharge, though concentrating wastes requiring alternative disposal.

Feed sustainability receives increasing attention as aquaculture expansion strains supplies of traditional feed ingredients, particularly fishmeal and fish oil from wild forage fish. Development of alternative protein sources including plant proteins, insect meals, single-cell proteins from bacteria or yeast, and eventually cellular agriculture products aim to reduce dependence on wild fish resources. Improved feed formulations utilize alternative ingredients while maintaining nutritional adequacy and growth performance. These efforts address legitimate concerns about aquaculture consuming more fish as feed than it produces as output—a situation true for some species but not others and diminishing as feed formulations evolve.

Certification and eco-labeling programs provide market-based incentives for sustainable production. Organizations including the Aquaculture Stewardship Council, Best Aquaculture Practices, and others establish standards for responsible aquaculture encompassing environmental impacts, social responsibility, animal welfare, and food safety. Certified products can access premium markets and potentially command higher prices, rewarding producers adopting best practices. Market pressure from major retailers and restaurant chains increasingly demands certified sources, accelerating adoption of sustainability standards.

Integrated multi-trophic aquaculture cultivates multiple species occupying different trophic levels in coordinated systems where wastes from one species provide nutrients for another. For example, fish culture effluent may fertilize seaweed production, while filter-feeding shellfish consume particulates from fish wastes. These integrated systems mimic natural ecosystem nutrient cycling, potentially improving overall sustainability and economic returns through diversified production.

Market Development and Value Chain Evolution

Aquaculture market dynamics are evolving rapidly as production scales up, supply chains modernize, and consumer preferences shift. Understanding these market trends is essential for aligning production with demand.

Market segmentation is increasing as consumers differentiate among products based on species, quality attributes, production methods, sustainability credentials, and convenience. Premium markets reward superior quality, freshness, sustainable production, and convenient product forms with price premiums, while commodity markets remain price-sensitive. Producers strategically target market segments matching their capabilities and cost structures.

Value addition through processing, branding, and marketing enhances returns beyond commodity prices for undifferentiated products. Filleting, portioning, marinating, smoking, and other processing creates convenience products commanding higher prices. Branding establishes product identity and consumer recognition, potentially building customer loyalty. Direct marketing through farmers markets, restaurants, or online sales captures retail margins while providing market intelligence and customer relationships.

Globalization of aquaculture trade continues expanding as transportation infrastructure improves, supply chains become more efficient, and trade barriers decline. Major producing nations export substantial production to international markets, particularly developed nations with high fish consumption but limited domestic production. Efficient global supply chains deliver fresh or frozen products to distant markets within hours or days of harvest, enabling international trade in products once consumed only locally.

Consumer preferences are evolving toward increased seafood consumption driven by health awareness, dietary diversification, and protein preferences. Simultaneously, consumers increasingly value sustainability, quality, food safety, and traceability. Health-conscious consumers appreciate omega-3 fatty acids and lean protein provided by fish. Concerns about mercury, antibiotics, or environmental impacts influence purchasing decisions. These evolving preferences create opportunities for producers meeting consumer expectations while challenging those failing to adapt.

Regional Development Patterns

While global trends provide overall context, regional patterns reflect specific conditions, constraints, and opportunities shaping aquaculture development in different parts of the world.

Asia maintains dominant position in global aquaculture but faces challenges including land and water scarcity, environmental degradation, disease pressures, and increasing production costs. Future growth will likely emphasize productivity improvement through genetic advances, better management, and disease control rather than extensive area expansion. Intensification continues, particularly for high-value species. Environmental regulations are strengthening in response to pollution concerns, requiring improved practices. The region will certainly maintain global leadership but growth rates may moderate.

Sub-Saharan Africa represents perhaps the greatest growth potential, given abundant land and water resources, growing population, protein deficit, and economic development creating markets. However, infrastructure limitations, technical capacity constraints, inadequate policy frameworks, and limited access to inputs and markets constrain rapid development. Successful African aquaculture development requires addressing these systemic constraints through infrastructure investment, capacity building, research adaptation to African conditions, and supportive policies. Species emphasis includes tilapias, African catfishes, and potentially indigenous species as culture technologies develop.

Latin America will likely continue salmon and shrimp focus while exploring diversification. Chilean salmon industry has matured into a global leader despite disease challenges requiring ongoing management. Shrimp production faces disease and environmental pressures but continues significant production. Native species including tambaqui and pirarucu offer diversification possibilities as culture technologies improve.

Europe and North America face limited expansion potential due to environmental regulations, high costs, and social resistance to facility siting. Production will likely emphasize high-value species, premium quality, sustainability credentials, and advanced technologies including recirculating systems. Aquaculture in these regions may never achieve the scale seen in Asia but can be economically viable serving demanding local markets willing to pay premium prices.

Climate change emerges as an overarching factor affecting all regions. Temperature changes may alter suitable production areas, species distributions, growing seasons, and disease dynamics. Water availability changes from altered precipitation patterns, glacial melt, or drought affect production possibilities. Ocean acidification and warming impact marine aquaculture. Adaptation strategies including species selection, facility relocation, and production system modification will be necessary as climate impacts intensify.

Policy and Institutional Development

Supportive policy and institutional frameworks are increasingly recognized as essential for sustainable aquaculture development. Progressive governments establish clear regulatory frameworks, provide infrastructure, invest in research and extension, facilitate access to credit, and promote market development.

Aquaculture policy must balance production promotion with environmental protection, coordinate multiple agencies with overlapping jurisdictions, and integrate with broader development strategies. Progressive policies establish clear procedures for site approval, environmental assessment, and operational permitting while providing reasonable timelines and predictable processes. Spatial planning identifies suitable aquaculture zones while protecting sensitive environments and minimizing conflicts with other water users.

Research and development investment is critical for technological advancement, problem-solving, and adaptation to changing conditions. Public research institutions, often in partnership with industry, develop improved culture practices, genetic strains, disease solutions, and management innovations that benefit the sector broadly. Extension services transfer research findings to practitioners, providing technical support and facilitating adoption of innovations.

Education and training programs build human capacity essential for sector development. University programs in aquaculture, fisheries science, and related disciplines provide professional-level education. Vocational training programs develop technical skills for farm workers and managers. Farmer training and extension educate practicing aquaculturists in improved methods.

Credit and financial services access enables farm establishment, expansion, and operational financing. Many potential farmers lack capital for investment or working capital for operational expenses. Development banks, microfinance institutions, and commercial banks with aquaculture expertise can fill this financing gap, enabling participation by creditworthy farmers lacking personal capital.

These diverse trends and developments paint a picture of a dynamic, rapidly evolving sector facing both opportunities and challenges. Successful navigation of this complex landscape requires understanding these trends, anticipating future directions, and adapting strategies accordingly. Pisciculture's continued growth and increasing importance in global food systems appear assured, while specific modalities, locations, and participants will certainly evolve substantially in coming decades.

Conclusion

This introductory chapter has established the foundational understanding essential for studying applied pisciculture as a scientific discipline and practical enterprise. We have defined pisciculture as the scientific rearing, breeding, and management of fish under controlled conditions, emphasizing its applied orientation toward productive, profitable, and sustainable commercial operations. The historical development from ancient subsistence practices to modern industrial-scale systems illustrates humanity's long engagement with fish culture and the technological sophistication achieved in recent decades.

The critical importance of pisciculture for global food security, nutritional adequacy, and economic development cannot be overstated. As capture fisheries stagnate and human population continues growing, aquaculture increasingly provides the fish protein consumed globally. The sector's contribution extends beyond simple tonnage to encompass high-quality nutrition, livelihood opportunities, rural development, and export earnings particularly benefiting developing nations.

Understanding the spectrum of culture systems from extensive through semi-intensive to intensive approaches provides essential context for appreciating how aquaculture adapts to diverse circumstances. Each system type offers particular advantages and limitations, with appropriate application depending on available resources, technical capabilities, market characteristics, and production objectives. The fundamental principles governing biological compatibility, environmental management, production operations, economic viability, and sustainability transcend specific systems or species, providing universal guidance for effective fish farming.

Global and regional trends reveal an industry experiencing rapid evolution driven by market demands, technological innovations, sustainability imperatives, and policy developments. The dramatic production expansion, particularly in Asia, demonstrates aquaculture's capacity for growth. Intensification trends, technological sophistication, genetic improvement, and sustainability emphasis characterize contemporary development. Regional patterns reflect diverse conditions and development trajectories, while policy and institutional frameworks increasingly influence development pathways.

The knowledge foundation established in this chapter prepares readers for detailed exploration of specific technical topics in subsequent chapters. Understanding fish biology, breeding technologies, hatchery management, water quality, nutrition, health management, and economic analysis requires building on these fundamental concepts. Applied pisciculture represents an integration of biological sciences, technology, management skill, and business acumen. Success requires mastering this interdisciplinary knowledge and skillfully applying it within the particular circumstances of individual operations.

As we proceed through subsequent chapters examining specific aspects of applied fish farming in detail, maintain perspective on how individual topics connect to the broader framework established here. Each technical component—whether breeding, feeding, disease management, or harvest logistics—ultimately serves the overarching goal of producing fish efficiently, profitably, and sustainably to meet human needs while preserving environmental integrity and social equity. This integration of knowledge toward practical application defines applied pisciculture and distinguishes it from purely academic study of fish biology.

CHAPTER 2

BIOLOGY AND BREEDING OF CULTIVABLE FISHES

Key Outcomes

Upon completing this chapter, readers will be able to:

1. Identify and apply criteria for selecting fish species suitable for culture in specific environmental and market contexts
2. Describe anatomical and physiological features of fish relevant to culture management and breeding
3. Explain growth patterns and factors affecting growth rates in cultivable fishes
4. Analyze reproductive biology including spawning requirements and behaviors of major carps and exotic species
5. Distinguish among natural, induced, and controlled breeding techniques and determine appropriate applications
6. Implement effective broodstock management programs ensuring availability of quality breeding material
7. Conduct spawning operations using appropriate techniques for different species
8. Establish and manage hatchery operations for efficient seed production

Abstract

This chapter provides comprehensive coverage of the biological foundations and breeding methodologies essential for successful fish culture operations. Understanding fish biology—including anatomy, physiology, growth patterns, and particularly reproductive biology—forms the prerequisite for effective breeding programs and hatchery management. The chapter systematically examines criteria and processes for selecting cultivable fish species, explores anatomical and physiological features relevant to culture, and analyzes growth dynamics affecting production planning. Detailed attention focuses on reproductive biology of major cultivable species including Indian major carps, Chinese carps, and exotic species like tilapia and catfishes. Natural, induced, and controlled breeding techniques are comprehensively described, emphasizing practical implementation suitable for different operational scales and technical capacities. Broodstock management principles covering selection, conditioning, handling, and maintenance ensure availability of quality breeding material. Spawning practices for different species are thoroughly examined, followed by detailed coverage of seed production processes and hatchery management systems. This chapter bridges fundamental biological knowledge with practical breeding applications, enabling readers to implement successful breeding programs supporting commercial fish production.

2.1 Selection of Cultivable Fish Species

The selection of appropriate fish species for culture represents perhaps the most fundamental decision affecting the ultimate success or failure of aquaculture ventures. While hundreds of fish species have been experimentally cultured, only a relatively small number have achieved commercial significance. Understanding the characteristics that make species suitable for culture and systematically evaluating candidate species against these criteria enables informed selection decisions matching species capabilities with environmental conditions, available resources, technical expertise, and market requirements.

Fundamental Requirements for Cultivable Species

Successful culture species must satisfy multiple biological, technical, and economic requirements simultaneously. Species lacking any critical requirement typically prove unsatisfactory for commercial production regardless of their other favorable characteristics.

Controlled reproduction under culture conditions represents the most fundamental requirement. Without reliable breeding and seed production, commercial culture remains impossible or dependent on unpredictable wild seed collection. Ideal culture species breed naturally in captivity when provided appropriate environmental conditions and management. Species requiring highly specific natural spawning conditions difficult to replicate in culture may breed successfully through induced spawning using hormonal preparations stimulating gamete maturation and spawning behavior. Species resistant to both natural and induced breeding methods under practical hatchery conditions generally prove unsuitable for widespread commercial culture, though exceptions exist where wild seed availability is reliable and sustainable.

Rapid growth under culture conditions is highly desirable, though the definition of rapid varies among species and culture contexts. Fast-growing species reach marketable size quickly, reducing production cycle duration, lowering cumulative costs, and enabling rapid capital turnover. Species requiring multiple years to achieve harvest size generally prove economically unfavorable except where very high market prices compensate for prolonged production periods. Comparison of growth rates should consider not absolute size but rather time required to reach market size for a particular target market.

Efficient feed conversion translates feed inputs into fish biomass economically. Feed typically constitutes the largest operational cost in intensive and semi-intensive culture, so efficient conversion directly affects profitability. Feed conversion ratio—the amount of feed required to produce a unit of fish growth—varies enormously among species, with lower ratios indicating more efficient conversion. Herbivorous and omnivorous species generally convert feed more efficiently than strict carnivores requiring high-protein diets. However, market price must be considered alongside feed conversion; high-value species may be economically viable despite less efficient feed conversion.

Disease resistance and general hardiness affect survival rates, production predictability, and health management costs. Hardy species tolerating environmental fluctuations, handling stress, and transport experience lower mortality and require less intensive health management. Disease-resistant species suffer fewer disease outbreaks, reducing treatment costs and improving reliability. Some species, while otherwise attractive for culture, prove highly susceptible to specific pathogens causing serious production losses. Disease resistance can sometimes be improved through selective breeding, but inherent susceptibility remains a significant selection consideration.

Environmental tolerance determines where and how species can be cultured. Wide-ranging tolerance for temperature, dissolved oxygen, pH, salinity, and water quality enables culture across diverse environments and reduces the sophistication of environmental control required. Stenothermal species with narrow temperature tolerance require temperature control limiting culture to specific geographical regions or necessitating expensive heating or cooling. Euryhaline species tolerating wide salinity ranges enable culture in brackish or even marine environments. Oxygen-tolerant species can be stocked at higher densities before dissolved oxygen becomes limiting.

Dietary habits and feeding behavior significantly affect culture management and economics. Herbivorous species can be raised on relatively inexpensive plant-based feeds or even natural productivity, minimizing feeding costs. Omnivorous species accept diverse feed types enabling dietary flexibility. Carnivorous species require expensive high-protein feeds, typically increasing production costs substantially. Feeding behavior also matters; species feeding throughout the water column utilize feeds more efficiently than bottom feeders potentially missing feed before it settles. Species accepting artificial feeds readily adapt better to intensive culture than those preferring natural live foods.

Market acceptability ultimately determines economic viability regardless of how well species satisfy biological and technical criteria. Target consumers must find the species acceptable based on taste, texture, appearance, cultural familiarity, and price point. Species unfamiliar to target markets may face consumer resistance regardless of nutritional value or production efficiency. Traditional food fishes generally enjoy immediate market acceptance, while novel species may require market development and consumer education. Product form also influences acceptability; some species perform better as whole fish while others are more acceptable as fillets or processed products.

Economic viability synthesizes all previous considerations, determining whether culture generates acceptable financial returns. Production costs including seed, feed, land, labor, and capital must be balanced against revenue at prevailing market prices. Species with favorable biological characteristics but low market value or high production costs may prove economically unviable. Conversely, high-value species may be profitable despite biological or technical challenges if revenues adequately exceed costs.

Process of Species Selection

Systematic species selection involves methodically evaluating candidate species against the requirements outlined above within the specific context of a proposed culture operation. This evaluation considers the unique combination of environmental conditions, available resources, technical capacity, target markets, and operational objectives characterizing each situation.

Environmental assessment begins by thoroughly characterizing available water resources including temperature regimes, dissolved oxygen levels, pH, alkalinity, salinity, and potential pollutants or water quality issues. Natural seasonal variation and extremes must be considered, not just average conditions. Available land characteristics including topography, soil type, and size determine facility design possibilities. Climate including temperature extremes, rainfall patterns, and severe weather risks affects species suitability and infrastructure requirements.

Species screening initially identifies all species potentially compatible with assessed environmental conditions. Reference materials including aquaculture textbooks, research literature, and regional aquaculture guides provide information on species environmental requirements and culture characteristics. Local and regional aquaculture practice offers valuable guidance; species successfully cultured in nearby areas with similar conditions are strong candidates. However, successful exotic species from distant regions may also warrant consideration if environmental compatibility can be verified.

Detailed evaluation of shortlisted candidate species examines biological characteristics, culture technologies, seed availability, market demand, economic feasibility, and risk factors. Biological characteristics are assessed including growth rates under conditions similar to the proposed operation, feed conversion efficiency, disease susceptibility, and handling tolerance. Culture technology requirements are evaluated against available technical capacity and willingness to develop necessary expertise. Seed availability through hatchery purchase, self-production, or wild collection affects production planning and risk. Market demand analysis examines price levels, seasonal patterns, product forms, market competition, and consumer preferences.

Economic analysis comparing candidate species provides quantitative comparison of expected financial performance. Production budgets estimate revenues and costs for each species given local conditions, producing profitability projections. Sensitivity analysis examines how profitability responds to variations in key parameters including market price, feed cost, survival rates, and growth rates. This analysis identifies which species offer best economic prospects and which variables most critically affect outcomes.

Risk assessment identifies major risks including biological uncertainties, market volatility, technical challenges, and external factors. Species with extensive commercial culture history and well-developed technologies generally present lower technical risk than newly developing species. Volatile market prices create financial risk, while stable markets offer greater predictability. Environmental risks including temperature extremes, water supply reliability, and disease occurrence vary among species and locations.

Final selection considers all evaluation dimensions holistically rather than optimizing any single criterion. The optimal species choice depends on whether objectives emphasize income generation, household food security, risk minimization, or other goals. In many cases, polyculture of multiple compatible species proves advantageous, balancing risks and utilizing environmental resources efficiently.

Major Cultivable Species Groups

While comprehensive description of all cultivable species exceeds this text's scope, understanding major species groups provides valuable context for subsequent detailed discussion of representative species biology and breeding.

Carps dominate global freshwater aquaculture production, particularly in Asia. Indian major carps including catla, rohu, and mrigal constitute traditional culture species in South Asia. Chinese carps including grass carp, silver carp, bighead carp, and common carp are cultured extensively worldwide. These species demonstrate favorable culture characteristics including relatively rapid growth, omnivorous or herbivorous feeding reducing feed costs, environmental tolerance, and strong market acceptance in Asian countries. Carp polyculture systems efficiently utilize pond resources by stocking compatible species with complementary feeding habits.

Tilapia species, particularly Nile tilapia, have become globally important food fishes. Tilapia demonstrates exceptional culture characteristics including rapid growth, high-density tolerance, acceptance of diverse feeds, disease resistance, and ease of breeding. Market acceptance has expanded globally beyond traditional African and Middle Eastern markets. However, prolific breeding can create management challenges in mixed-sex culture, motivating development of sex reversal techniques producing all-male populations growing faster than females.

Catfishes are cultured extensively in various regions with different species dominating locally. Channel catfish dominates North American aquaculture, while African catfish and Asian catfish species are important elsewhere. Catfish generally demonstrate good growth, environmental tolerance, and market acceptance where culturally familiar. Most catfish accept prepared feeds readily. Some species exhibit air-breathing abilities enabling survival in low-oxygen conditions and very high-density culture.

Coloration results from specialized pigment cells called chromatophores located in skin and scales. Fish coloration serves diverse functions including camouflage, species recognition, and communication. Color patterns often differ between males and females, particularly during breeding seasons. In culture, coloration may reflect health status, with diseased or stressed fish sometimes exhibiting darkening or abnormal patterns. Market acceptability sometimes relates to coloration, affecting species selection and breeding objectives.

Lateral line system is a specialized sensory structure unique to fishes and some aquatic amphibians, appearing externally as a line of modified scales along each body side. This mechanosensory system detects water movements and pressure changes, enabling detection of obstacles, predators, prey, and conspecifics even in murky water or darkness. The lateral line's functionality is important for normal behavior and feeding.

Respiratory System and Gas Exchange

Fish extract dissolved oxygen from water through highly efficient gill structures, with respiratory efficiency fundamentally affecting culture density limitations and management requirements.

Gill structure consists of paired gill arches on each side of the head, protected by gill covers called opercula. Each arch bears numerous gill filaments, delicate finger-like projections oriented perpendicular to the arch. Each filament in turn bears hundreds of microscopic lamellae, tiny plates arranged perpendicular to the filament. This intricate structural arrangement creates enormous surface area for gas exchange relative to body size.

Gas exchange occurs as water flows over gill lamellae, bringing dissolved oxygen into intimate contact with blood flowing through extensive capillary networks within lamellae. Oxygen diffuses from water into blood along concentration gradients, while carbon dioxide diffuses from blood into water. The countercurrent flow arrangement, where blood and water flow in opposite directions through gill lamellae, maximizes gas exchange efficiency by maintaining favorable concentration gradients along the entire lamella length.

Oxygen requirements vary substantially among species, life stages, temperatures, activity levels, and feeding status. Warm-water species generally tolerate lower dissolved oxygen levels than cold-water species, reflecting different metabolic adaptations. Small fish have relatively higher oxygen requirements per unit body weight than large fish due to higher metabolic rates. Temperature profoundly affects oxygen requirements, with requirements roughly doubling for each 10 degree Celsius temperature increase within species tolerance ranges. Fish are most vulnerable to low oxygen several hours after feeding when metabolic oxygen demand peaks.

Air-breathing adaptations enable some species to survive in oxygen-depleted waters or even temporary drying of habitats. Various air-breathing organs including labyrinth organs, modified swim bladders, or vascularized mouth and pharynx tissues enable atmospheric oxygen uptake. Air-breathing species including snakeheads, walking catfish, and some gouramis tolerate conditions lethal to species dependent solely on gill respiration. This adaptation enables higher stocking densities in culture and reduces aeration requirements.

Digestive System and Nutrient Processing

Fish digestive systems exhibit remarkable diversity reflecting varied dietary habits, with digestive anatomy and physiology profoundly influencing feeding management and nutrition in culture.

Mouth and oral cavity structure varies extensively among species with different feeding modes. Carnivorous species typically possess large terminal or superior mouths with numerous sharp teeth for grasping and tearing prey. Herbivorous species often have smaller inferior mouths with specialized dentition for scraping or grinding plant material. Filter feeders may possess small mouths with specialized gill rakers straining plankton from water pumped through the mouth and gill apparatus. Some species possess pharyngeal teeth located in the throat region grinding ingested food.

Digestive tract anatomy reflects dietary habits, with carnivores typically possessing short, straight digestive tracts while herbivores have longer, more convoluted intestines accommodating extended digestion times required for plant materials. The digestive tract consists of esophagus conducting food from mouth to stomach, stomach where initial digestion occurs through acid and enzyme secretion, intestines where most digestion and nutrient absorption occur, and rectum conducting waste to the anus.

Stomach presence and structure varies among species. Many species possess distinct, expandable stomachs where ingested food is stored and partial digestion occurs. Other species lack distinct stomachs, with the esophagus opening directly into intestines. Species with stomachs can consume large meals intermittently, while stomachless species typically require more frequent smaller feedings.

Liver and pancreas produce digestive enzymes and other secretions essential for digestion. The liver, the largest internal organ in most fishes, produces bile salts emulsifying dietary lipids, stores glycogen and other nutrients, and performs numerous metabolic functions. The pancreas secretes digestive enzymes including proteases digesting protein, lipases digesting fats, and amylases digesting carbohydrates.

Nutrient absorption occurs primarily in the intestines through specialized absorptive cells lining the intestinal wall. Intestinal length correlates generally with diet type, with herbivores possessing longer intestines increasing surface area and time for digesting plant materials. Carnivorous species with shorter intestines efficiently digest and absorb animal tissues requiring less extensive processing.

Digestive efficiency affects feed conversion rates and production economics. Efficient digesters extract more nutrients from consumed feed, growing faster per unit feed consumed. Digestive efficiency varies with diet composition, feed quality, temperature, fish size, and feeding frequency. Understanding these relationships enables optimization of feeding practices maximizing growth while minimizing feed waste.

Excretory System and Osmoregulation

Fish excretory systems eliminate metabolic wastes while maintaining proper water and ion balance. Understanding excretion and osmoregulation is essential for water quality management in culture systems.

Kidneys filter blood, removing nitrogenous wastes and excess water or ions depending on environmental salinity. Fish kidneys function somewhat differently than mammalian kidneys due to the challenges of living in aquatic environments with varying salinity. Freshwater fish excrete copious dilute urine eliminating excess water constantly entering by osmosis while conserving salts. Marine fish excrete small volumes of concentrated urine, conserving water while eliminating excess salts entering from the surrounding environment.

Ammonia excretion through gills constitutes the primary route for nitrogenous waste elimination in most fishes. Fish are ammonotelic, excreting ammonia directly rather than converting it to urea or uric acid as terrestrial vertebrates must. Ammonia excretion is efficient but requires ample water for dilution, since ammonia becomes toxic at low concentrations. In culture systems, ammonia accumulates in water, necessitating management through water exchange, aeration, or biological filtration.

Osmoregulation maintains proper internal salt and water balance despite living in environments very different in concentration from body fluids. Freshwater fish are hypertonic relative to surrounding water, constantly gaining water by osmosis and losing salts by diffusion. To compensate, freshwater fish produce copious dilute urine eliminating excess water while specialized gill cells actively absorb salts from water. Marine fish are hypotonic relative to seawater, constantly losing water by osmosis and gaining salts by diffusion. Marine fish compensate by drinking seawater, absorbing water while excreting excess salts through specialized gill cells and concentrated urine. These osmoregulatory processes consume substantial energy, affecting growth rates.

Salinity tolerance varies enormously among species. **Stenohaline** species tolerate only narrow salinity ranges, restricted to either freshwater or saltwater environments. **Euryhaline** species tolerate wide salinity variations, enabling them to inhabit brackish waters or migrate between freshwater and marine environments. Some euryhaline species including tilapia and striped bass can be cultured across wide salinity ranges, providing operational flexibility. Gradual salinity acclimation allows euryhaline species to adjust osmoregulatory mechanisms, while sudden salinity changes can cause severe stress or mortality.

Growth Patterns and Factors Affecting Growth

Fish growth—increase in size and weight over time—represents the fundamental production process in aquaculture, with growth rate directly determining production cycle duration and efficiency. Understanding factors controlling growth enables management optimization.

Growth basics reflect that growth occurs when energy and nutrients obtained from food exceed those expended for maintenance metabolism, activity, and other processes. Surplus energy and nutrients are allocated to tissue synthesis, increasing body size and weight. The fundamental growth equation conceptually states that growth equals consumption minus metabolism minus excretion. Maximizing growth requires optimizing food consumption while minimizing unnecessary metabolic expenditure.

Temperature profoundly affects growth rates in fish as cold-blooded organisms whose metabolic rates are strongly temperature-dependent. Each species exhibits an optimal temperature range where growth is maximized. Below this range, metabolic rate and growth slow progressively as temperature decreases. Above optimal temperatures, metabolic costs increase faster than feeding and growth rates, with net growth declining. At extreme high temperatures approaching lethal limits, feeding may cease entirely. Understanding species-specific temperature-growth relationships enables timing production to utilize favorable temperature periods and selecting species matched to local temperature regimes.

Food quality and quantity fundamentally determine growth rates given appropriate environmental conditions. Nutritionally complete diets supporting maximum growth provide optimal ratios of protein, carbohydrates, lipids, vitamins, minerals, and essential nutrients. Imbalanced or deficient diets cause reduced growth even if feeding levels are adequate. Feed quantity must match fish appetite and digestive capacity; underfeeding limits growth while overfeeding wastes feed and degrades water quality without additional growth. Optimal feeding strategies adjust ration size and frequency based on fish size, temperature, and observed feeding response.

Stocking density affects growth through multiple mechanisms including competition for food and space, behavioral interactions, and water quality impacts. At low densities where resources are abundant and water quality remains excellent, individual fish may approach their genetic growth potential. As density increases, individual growth rates typically decline due to increased competition, stress, or deteriorating water quality even if feeding and oxygen appear adequate. Understanding density-growth relationships enables balancing individual growth rates against total biomass production per unit culture volume.

Fish size influences growth rates with specific growth rates generally declining as fish age and approach adult size. Small fish grow rapidly on a percentage basis, potentially doubling body weight in days or weeks. Large fish growing in absolute terms may actually be adding more grams daily than small fish but represent smaller percentage increases. This size-dependent growth affects production planning, with fastest weight gain often occurring during middle portions of production cycles rather than at beginning or end.

Genetic variation in growth potential exists within species, with some individuals or families exhibiting substantially faster growth than others under identical conditions. This genetic variation provides raw material for selective breeding programs improving growth rates over generations. Genetically improved strains may grow 20 to 80 percent faster than wild populations or unselected culture stocks, dramatically improving production efficiency. Understanding genetic influences on growth motivates maintaining quality broodstock and potentially participating in genetic improvement programs.

Sex differences in growth occur in many species, with males and females exhibiting different growth patterns. In tilapia, males grow substantially faster than females, motivating production of all-male populations through hormone treatment or genetic techniques. In some species, females grow faster or larger than males. Understanding sex-specific growth patterns may influence culture strategies.

Growth measurement and monitoring in culture operations provides essential feedback on production progress and enables adaptive management. Periodic sampling, weighing, and measuring representative fish generates growth data used to calculate specific growth rates, evaluate feeding programs, predict harvest timing, and identify problems. Weight typically provides more useful information than length for production purposes, though length is easier to measure nondestructively. Condition factor—the relationship between weight and length—provides insight into fish nutritional status and appropriateness of body form.

Table 2: Growth-Related Factors and Management Implications

Factor	**Effect on Growth**	**Management Strategy**
Temperature	Optimal range for maximum growth	Select species suited to climate; time production
Feed Quality	Nutritionally complete feeds support best growth	Use quality feeds formulated for species
Feed Quantity	Sufficient but not excessive	Adjust feeding to appetite and temperature
Dissolved Oxygen	Adequate oxygen essential; low levels reduce growth	Aerate ponds; manage stocking density
Stocking Density	Higher density often reduces individual growth	Balance density against individual growth
Fish Health	Disease reduces or halts growth	Prevent disease through biosecurity and optimal conditions
Water Quality	Poor quality suppresses growth	Monitor and maintain ammonia, pH, alkalinity
Genetics	Superior strains grow faster	Use improved strains; maintain quality broodstock

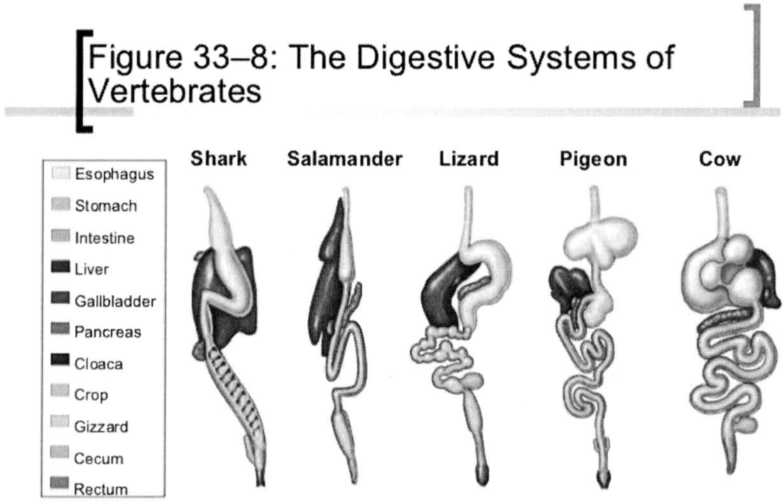

Figure. 4: Comparative Digestive Tract Anatomy

2.3 Reproductive Biology of Major Carps and Exotic Fishes

Successful breeding and seed production require thorough understanding of reproductive biology including sexual maturity, spawning requirements, breeding behavior, and gamete characteristics. This section examines reproductive biology of major cultivable species, focusing on details most relevant to breeding programs and hatchery operations.

Sexual Maturity and Breeding Seasonality

Fish sexual maturity varies substantially among species and is influenced by factors including genetics, environmental conditions, nutrition, and population density. Understanding maturity patterns enables appropriate broodstock selection and breeding program timing.

Age and size at maturity vary enormously, from several months in small rapidly-reproducing species to several years in larger species. Carps typically reach sexual maturity at ages of one to three years depending on species, growth rate, and environmental conditions. Tilapia can mature remarkably quickly, with some individuals becoming sexually mature at ages of three to six months when only weighing tens of grams. Salmonids generally require two to four years to mature. Size often relates more directly to maturity than age, with fish reaching specific body sizes triggering maturation regardless of chronological age. Rapid growth from optimal nutrition and environmental conditions can accelerate maturity.

Breeding seasonality characterizes most temperate and many tropical species, with spawning concentrated during specific seasons when environmental conditions favor larval survival. Environmental cues triggering breeding readiness include water temperature changes, photoperiod alterations, rainfall and water level fluctuations, and lunar cycles. Temperature commonly serves as a primary breeding cue, with species spawning when temperatures reach appropriate ranges after cool periods. In tropical species without strong temperature seasonality, photoperiod or rainfall may serve as primary cues.

Indian major carps including catla, rohu, and mrigal naturally spawn during southwest monsoon season from June through September when rising water levels, increased current, reduced water temperature, and turbidity created by monsoon rains trigger spawning behavior. In natural rivers and floodplains, these environmental changes stimulate simultaneous spawning across populations, with fertilized eggs drifting downstream before hatching. The specific environmental cues triggering spawning are remarkably consistent across the species group, suggesting shared evolutionary adaptation to monsoon-driven riverine breeding.

Chinese carps exhibit similar monsoon-associated breeding patterns in their native river habitats, spawning during rising water periods accompanying spring and summer rains. Grass carp, silver carp, bighead carp, and common carp require similar environmental cues, though optimal temperature ranges vary somewhat among species.

Tilapia breeding differs fundamentally from carps, showing little seasonality and breeding repeatedly throughout the year when conditions are favorable. Sexual maturity comes early, breeding frequency is high, and egg numbers per spawning are relatively modest compared to carps. This reproductive pattern reflects tilapia evolutionary origins in stable tropical lakes rather than seasonal rivers. While continuous breeding facilitates seed production, it creates management challenges in grow-out ponds where uncontrolled reproduction produces populations of mixed sizes.

Catfishes exhibit varied reproductive patterns depending on species. African catfish breeds seasonally in nature but can be induced to spawn throughout the year in hatcheries using hormonal methods. Asian catfishes similarly show seasonal natural breeding but hatchery production throughout the year.

Gonadal Development and Maturation

Understanding gonadal development cycles enables assessment of broodstock readiness and timing of breeding interventions.

Gonad structure differs fundamentally between sexes. Male gonads, the testes or milt glands, produce sperm and testicular fluid. Female gonads, the ovaries, produce eggs or ova. Both gonad types undergo cyclical development synchronized with breeding seasons in seasonal spawners.

Ovarian development proceeds through characteristic stages observable through external examination and internal sampling. Early developing ovaries are small, occupying only a small proportion of body cavity, with individual eggs microscopic and not distinguishable. Partially developed ovaries increase in size and become more vascular, with individual eggs becoming visible as tiny spheres but still translucent. Well-developed ovaries occupy substantial body cavity space, appearing as paired, dense, grainy structures with opaque yellowish or amber eggs easily visible. Fully mature ovaries ready for spawning are greatly enlarged, occupying most of the body cavity, with large, translucent eggs separating easily from ovarian tissue. In some species, mature eggs are released into the body cavity before spawning.

External indicators of female maturity include swollen, soft abdomens, reddened and slightly swollen genital openings, and general robust appearance. Gentle pressure on the abdomen of ripe females may cause release of a few eggs. However, external assessment alone can be unreliable; examining eggs by making a small incision or using cannulation techniques provides more definitive assessment. Mature eggs are large, round, uniform, and translucent, while immature eggs are smaller, opaque, and irregular.

Testicular development similarly proceeds through stages from immature to fully developed. Mature males produce copious milt—a white or cream-colored fluid containing millions of sperm cells and seminal fluids. Gentle abdominal pressure on ripe males causes milt expression from the genital opening, providing definitive evidence of maturity.

Hormonal control of gonadal maturation involves complex interactions among environmental factors, brain perception and processing of environmental cues, hormonal signals from the brain's hypothalamus and pituitary gland, and gonadal responses to hormonal stimulation. The hypothalamus produces gonadotropin-releasing hormone in response to appropriate environmental cues. This hormone stimulates the pituitary gland to produce and release gonadotropins—luteinizing hormone and follicle-stimulating hormone—into the bloodstream. Gonadotropins circulating to gonads stimulate final gamete maturation and sex steroid hormone production. This neuroendocrine axis provides targets for induced breeding interventions using exogenous hormones or hormonal preparations.

Spawning Behavior and Ecology

Fish exhibit remarkable diversity in spawning behaviors and ecology, with these behaviors influencing breeding management approaches.

Indian major carps are broadcast spawners releasing eggs and sperm into flowing water where external fertilization occurs. Natural spawning occurs in rivers during monsoon floods, with breeding fish migrating upstream to specific spawning areas. Males and females engage in characteristic courtship and spawning behavior including close following, body contact, and synchronized gamete release. The fertilized eggs are semi-buoyant, drifting downstream with current while developing. Hatchery breeding attempts to replicate key environmental stimuli triggering spawning or bypasses natural spawning through induced breeding.

Common carp spawning occurs in shallow, vegetated waters where females deposit adhesive eggs on submerged vegetation while males release milt. This spawning substrate requirement can be accommodated in hatchery breeding by providing artificial spawning substrates including fibrous materials or aquatic plants. Spawning is often riotous and splashing, with multiple males pursuing females.

Tilapia exhibit fascinating parental care behavior contrasting sharply with carp broadcast spawning. Most cultured tilapia species are mouthbrooders, with females taking fertilized eggs into their mouths where they are held during embryonic development. Breeding behavior includes establishment of breeding territories by males, often by constructing circular nests scooped from substrate. Females attracted to male territories spawn there, with males fertilizing eggs before females collect eggs into their mouths. Females carry developing eggs and newly hatched larvae in their buccal cavity for approximately two weeks, providing protection during vulnerable early life stages. This parental care increases offspring survival compared to broadcast spawning but reduces fecundity per spawning event.

African catfish naturally spawn in shallow flooded areas during rainy seasons, with males and females engaging in spawning embraces where they lie side-by-side releasing gametes simultaneously. The eggs are demersal and adhesive, sticking to substrate or vegetation. In hatchery breeding, artificial substrates can accommodate natural spawning behavior, though induced breeding methods are more commonly employed for reliable seed production.

Channel catfish spawn in cavities or protected spaces, with males establishing and defending nests. Females deposit egg masses in nests where males fertilize and subsequently guard them, fanning with fins to maintain water circulation and oxygen supply. This paternal care behavior enables collection of fertilized egg masses for artificial incubation in hatcheries.

Fecundity and Egg Characteristics

Understanding egg production and characteristics enables estimation of seed production potential and appropriate handling of eggs during hatchery operations.

Fecundity—the number of eggs produced per spawning—varies enormously among species and individuals, generally correlating with body size. Carps are highly fecund, with large females potentially producing hundreds of thousands to over a million eggs per spawning. A one-kilogram rohu female may produce 100,000 to 200,000 eggs, with fecundity increasing roughly proportionally with body weight. Catla and mrigal show similar relationships between body size and egg production. Chinese carps exhibit comparable high fecundity appropriate for riverine broadcast spawning strategies.

Tilapia fecundity is substantially lower than carps, reflecting different reproductive strategies. A 200-gram Nile tilapia female may produce 200 to 400 eggs per spawning, far fewer than similarly-sized carps. However, tilapia compensate through frequent spawning and parental care increasing offspring survival rates. When females can spawn every four to six weeks throughout the year, annual egg production can be substantial despite low per-spawning fecundity.

Catfish fecundity falls between tilapia and carps, with egg production of several thousand to tens of thousands per spawning depending on species and female size. African catfish females may produce 20,000 to 150,000 eggs per kilogram body weight, enabling significant seed production from relatively few broodstock.

Egg size varies among species, generally inversely related to fecundity. Species producing many small eggs invest less per egg but rely on quantity for reproductive success. Species producing fewer larger eggs invest more per offspring, often accompanying parental care. Carp eggs measure approximately 1 to 2 millimeters in diameter when mature, expanding somewhat upon water absorption after fertilization. Tilapia eggs are larger, measuring 2 to 3 millimeters, containing more yolk supporting longer embryonic development and larger hatchlings. Catfish eggs vary among species but generally measure 1 to 2 millimeters.

Egg structure comprises several components supporting embryonic development. The **yolk** provides nutritional reserves sustaining the embryo through development until feeding commences. Larger eggs contain more yolk enabling longer pre-feeding developmental periods. The **cytoplasm** contains cellular machinery directing development. The **egg membrane** or chorion surrounds the egg, providing structural integrity and protection while allowing gas exchange. Some species' eggs have **adhesive surfaces** causing eggs to stick to substrate or each other, while others are non-adhesive, free-floating, or semi-buoyant. Understanding egg characteristics influences handling procedures during hatchery operations.

Egg quality profoundly affects fertilization success, normal development, and hatching rates. High-quality eggs are large, uniform in size, translucent, and separate easily from ovarian tissue when mature. Poor-quality eggs may be small, irregular, opaque, or mixed with immature or over-ripe eggs. Egg quality depends on female nutritional condition, genetic background, environmental conditions during maturation, and handling care. Maintaining excellent broodstock conditions and proper handling techniques ensures high egg quality.

Sperm characteristics similarly affect fertilization success. Healthy sperm are highly motile upon contact with water, swimming vigorously for brief periods sufficient to locate and fertilize eggs. Sperm viability is brief once exposed to water, typically remaining motile for only seconds to minutes depending on species and environmental conditions. Milt quality depends on male nutritional status, maturation conditions, and handling. Fresh milt appears creamy white or pale yellow, flows freely, and exhibits vigorous sperm motility under microscopic examination.

Fertilization and Embryonic Development

Successful fertilization and normal embryonic development require proper gamete handling and environmental conditions during critical early life stages.

Fertilization process in broadcast spawners occurs externally when eggs and sperm are released simultaneously into water. Sperm swim to eggs, with successful sperm penetrating the egg membrane through specialized structures. The egg membrane prevents additional sperm entry after successful fertilization, preventing polyspermy that would cause abnormal development. In mouth-brooding tilapia, fertilization occurs just before or as the female collects eggs, ensuring fertilization timing despite parental care.

Successful fertilization requires several conditions being met simultaneously. Eggs must be fully mature and recently released, with fertilization success declining rapidly in over-ripe eggs. Sperm must be viable with good motility. Appropriate water quality including proper temperature, pH, and freedom from contamination is essential. The ratio of sperm to eggs must be adequate, ensuring sufficient sperm are present to fertilize all eggs. In artificial propagation, these conditions are carefully managed to maximize fertilization rates.

Embryonic development following successful fertilization proceeds through characteristic stages observable under magnification. Within hours, cell division commences, with the single fertilized cell dividing into two cells, then four, eight, and so forth in geometric progression. This cleavage stage produces increasing numbers of smaller cells. The **blastula stage** follows when the mass of dividing cells forms a hollow ball. **Gastrulation** reorganizes cells into distinct layers forming the basic body plan. Subsequent development forms recognizable structures including eyes, heart, and somites—paired segments visible along the body axis. The embryo gradually assumes fish-like appearance with distinct head, trunk, and tail regions, visible eyes, and developing fins. Throughout embryonic development, the yolk sac provides nutrition, with yolk gradually consumed as the embryo develops.

Environmental requirements during embryonic development are critical for normal development and survival. Temperature profoundly affects developmental rate, with development proceeding faster at higher temperatures within the species tolerance range. Excessively high temperatures cause abnormal development and death, while excessively low temperatures slow development and may cause mortality. Each species exhibits optimal temperature ranges supporting best survival and development. Dissolved oxygen must remain adequate for embryonic respiration, though oxygen requirements are initially modest and increase as development proceeds. Water quality including pH, alkalinity, and freedom from pollutants affects development success. Gentle water movement is beneficial for carps and some other species, providing oxygen and preventing fungal growth, but excessive turbulence can damage delicate eggs.

Hatching occurs when embryonic development completes and the larvae break free from egg membranes. Hatching time depends on species and temperature, ranging from approximately 24 hours in some warm-water species at high temperatures to several days at cool temperatures. Newly hatched larvae are tiny, delicate organisms initially dependent on yolk reserves for nutrition. The **yolk sac larvae**

stage follows hatching, during which the attached yolk sac continues providing nutrition while larval structures complete development. As yolk absorbs, larvae begin feeding on external food sources, transitioning to the **fry** stage. This transition represents a critical period requiring appropriate food availability and environmental conditions for survival.

Abnormal development sometimes occurs due to genetic defects, poor egg or sperm quality, improper environmental conditions during development, or physical damage to eggs. Common abnormalities include spinal deformities, malformed heads or mouths, failure of the swim bladder to inflate properly, and improper development of fins or eyes. Severely abnormal individuals typically die during embryonic development or soon after hatching. Mildly abnormal individuals may survive but exhibit reduced growth, impaired function, or poor market quality. High rates of abnormality indicate problems with broodstock quality, gamete handling, or incubation conditions requiring correction.

2.4 Breeding Techniques – Natural, Induced, and Controlled Breeding

Reliable breeding and seed production constitute fundamental requirements for successful fish culture. While some species breed readily in captivity given appropriate conditions, others require significant intervention to spawn reliably. This section examines the spectrum of breeding techniques from natural spawning through increasingly sophisticated induced and controlled breeding methods.

Natural Breeding in Captivity

Natural breeding involves creating conditions enabling fish to spawn spontaneously without direct manipulation or hormone administration. Species amenable to natural breeding in captivity demonstrate reproductive success when provided appropriate environmental conditions, substrate, space, and social settings.

Common carp breeds naturally in ponds when appropriate spawning substrate is provided during warm months. Breeding ponds of modest size with shallow areas and abundant vegetation or artificial substrates support natural spawning. Mature broodstock are stocked at appropriate ratios, typically two to three males per female, and monitored for spawning activity. Rising temperatures in spring or early summer trigger spawning behavior, with fish actively pursuing partners and releasing gametes over substrates. Adhesive eggs stick to vegetation or artificial substrates where they develop. After spawning, broodstock are removed to prevent predation on eggs and newly hatched fry. Egg-laden substrates can be transferred to hatching facilities or left in spawning ponds for natural hatching.

Tilapia demonstrate perhaps the easiest natural breeding among major cultivable species, spawning continuously throughout the year when water temperatures exceed approximately 22 to 24 degrees Celsius. No special facilities or interventions are required; mixed-sex populations maintained in ponds or tanks will breed naturally. Males establish territories and construct nests, females spawn with dominant males, and oral incubation ensures good survival of eggs and larvae. While natural breeding ease is advantageous for seed production, it creates management issues in grow-out culture through uncontrolled reproduction and recruitment.

For tilapia seed production, separate breeding ponds or tanks maintain broodstock at ratios of approximately three females per male. Females are periodically checked for egg presence by gently opening their mouths. Gravid females carrying developing eggs can be isolated in separate containers where they continue brooding until larvae are released. Alternatively, eggs can be carefully removed from females' mouths and transferred to hatching jars or similar facilities for artificial incubation, allowing females to recover more quickly and resume spawning.

Channel catfish natural breeding requires nest cavities where males establish territories and attract females. Spawning containers including milk cans, clay pipes, or wooden boxes placed in ponds provide suitable nest sites. Males guard nests and subsequently guard and aerate egg masses after spawning. Regular checking of nests reveals egg masses that can be transferred to hatchery incubation systems, or alternatively left with guarding males until hatching.

Advantages of natural breeding include simplicity, lower technical requirements, no need for hormones or specialized equipment, and potentially more natural timing synchronized with optimal environmental conditions. Species exhibiting reliable natural spawning in culture conditions simplify breeding operations considerably.

Limitations include dependence on environmental conditions, potential unpredictability in spawn timing, possible mismatch between breeding season and desired seed production periods, and limited control over genetic combinations. Additionally, major cultivable species including Indian major carps and Chinese carps generally do not spawn naturally in confined culture conditions, necessitating induced breeding techniques.

Induced Breeding Using Hormonal Preparations

Many economically important species including major carps do not spawn naturally in ponds or tanks despite reaching sexual maturity and developing mature gametes. These species evolved in large river systems where specific environmental cues associated with monsoon floods trigger spawning. Replicating these complex environmental conditions is difficult or impossible in hatchery settings. **Induced breeding** techniques overcome this barrier by administering hormones or hormone-containing preparations that stimulate final gamete maturation and spawning, bypassing environmental cue requirements.

Biological basis of induced breeding relates to the neuroendocrine control of reproduction. Natural environmental cues trigger the brain's hypothalamus to secrete gonadotropin-releasing hormone, which stimulates the pituitary gland to release gonadotropins. These gonadotropins induce final egg maturation, sperm maturation, sex steroid production, and spawning behavior. In species not receiving appropriate environmental cues in captivity, this hormonal cascade does not occur despite gonadal development. Induced breeding administers hormones or hormone-containing materials that either replace or stimulate production of gonadotropins, thereby completing the maturation process and inducing spawning.

Pituitary gland extracts were the original and remain widely used materials for induced breeding. The pituitary gland from mature fish of the same or related species contains high concentrations of gonadotropins. Collected pituitary glands are processed into injectable extracts that are administered to broodstock, providing gonadotropins that complete maturation and induce spawning. The Indian scientist H.C. Chaudhuri and colleagues pioneered induced breeding of Indian major carps in the 1950s using carp pituitary extracts, revolutionizing carp culture in Asia. This technique enabled reliable seed production supporting massive expansion of carp aquaculture.

Pituitary collection and preparation involves sacrificing mature donor fish, typically common carp, and removing the tiny pituitary gland located at the base of the brain. The gland is extracted carefully, cleaned of blood and tissue, and preserved by drying or acetone treatment. Dried pituitaries can be stored for extended periods. When needed for breeding, preserved pituitaries are weighed and ground or homogenized with sterile saline or distilled water to create an injectable suspension. The suspension is administered to broodstock through intramuscular injection, typically in the dorsal musculature near the base of the dorsal fin.

Dosage determination depends on species, maturity stage, environmental conditions, and pituitary potency. Typical doses for carps range from 2 to 8 milligrams of pituitary preparation per kilogram of female body weight, with males receiving approximately half the female dose. Breeding protocols often employ split dosing, administering an initial priming dose of approximately 20 to 30 percent of total dose followed by a resolving dose approximately six to eight hours later. This split dosing mimics more natural patterns of gonadotropin elevation.

Synthetic hormones including human chorionic gonadotropin, gonadotropin-releasing hormone analogs, and dopamine antagonists provide alternatives to pituitary extracts. These synthetic preparations offer advantages including standardized potency, elimination of donor fish sacrifice, longer storage life, and potentially more consistent results. Human chorionic gonadotropin is extracted from human pregnancy urine and possesses gonadotropin activity effective in fish. Commercially available formulations are widely used in fish breeding. Gonadotropin-releasing hormone analogs including synthetic peptides directly stimulate the fish's own pituitary to release gonadotropins. These are often combined with dopamine antagonists that block dopamine inhibition of gonadotropin release, enhancing effectiveness.

Ovaprim represents a commercially available synthetic hormone preparation widely used for induced breeding. It combines a gonadotropin-releasing hormone analog with domperidone, a dopamine antagonist, in a ready-to-inject solution. Ovaprim is administered at much lower volumes than pituitary extracts, typically 0.5 to 1.0 milliliter per kilogram body weight for females. Its standardized formulation, ease of use, and consistent results have made it popular despite higher per-dose costs compared to pituitary extracts.

Breeding procedure following hormone injection requires careful management. Injected broodstock are transferred to spawning facilities including breeding hapas—fine-meshed cloth enclosures suspended in ponds or tanks—or small breeding pools with flowing water. Male and female broodstock are placed together at ratios typically of two males per female, though ratios vary among protocols. Water circulation through breeding containers provides current stimulating spawning behavior. Following the latency period between injection and spawning—typically 6 to 12 hours depending on temperature—broodstock exhibit characteristic breeding behavior including chasing, close following, and body contact. Spawning occurs during early morning hours in most cases, with fertilized eggs released into water.

Stripping technique represents an alternative to allowing natural spawning after hormone injection. When females show clear signs of egg release readiness including freely flowing eggs when gentle abdominal pressure is applied, eggs can be stripped manually into dry containers. Males are similarly stripped, with milt collected in separate containers. Eggs and milt are mixed thoroughly but gently, with a small amount of water added to activate sperm motility and initiate fertilization. After allowing several minutes for fertilization, eggs are rinsed and transferred to incubation facilities. Stripping provides more control over fertilization, enables specific breeding combinations, prevents egg loss to predation, and allows precise timing. However, it requires greater technical skill and can damage fish if performed improperly.

Post-spawning management involves removing broodstock from breeding containers after spawning or stripping is complete. Broodstock should be handled carefully as they are exhausted and vulnerable following breeding. They are returned to holding facilities where good water quality and minimal stress support recovery. Fertilized eggs collected from breeding containers or produced through stripping are transferred immediately to hatching facilities for incubation.

Controlled Breeding and Genetic Management

Controlled breeding refers to deliberate selection and mating of specific individuals to achieve genetic improvement, maintain genetic diversity, prevent inbreeding, or produce specific hybrid combinations. Controlled breeding integrates selective breeding principles with practical breeding techniques.

Selective breeding programs aim to improve economically important traits including growth rate, feed conversion efficiency, disease resistance, environmental tolerance, product quality, or reproductive characteristics. Programs begin by establishing base populations with genetic diversity. Individual fish are performance-tested, measuring traits of interest. Superior performers are selected as breeding candidates. Selected individuals are bred, producing the next generation that undergoes similar evaluation and selection. This cyclical process, repeated over multiple generations, accumulates favorable genes and improves trait values.

Successful selective breeding requires maintaining adequate population size preventing inbreeding, accurate performance measurement enabling reliable selection, sufficient selection intensity choosing the best individuals, and controlled mating ensuring superior individuals actually become parents of the next generation. Record-keeping documenting parentage, performance, and breeding history is essential. Modern selective breeding may incorporate **genetic markers** identifying individuals carrying favorable genes, enabling more efficient selection.

The **GIFT program** for Nile tilapia exemplifies successful selective breeding in aquaculture. This international collaboration initiated in the 1980s collected diverse tilapia strains, established base populations, implemented rigorous performance testing and selection over multiple generations, and produced genetically improved strains growing 60 to 80 percent faster than wild-type tilapia. GIFT tilapia have been disseminated globally, substantially improving tilapia production efficiency.

Hybridization involves mating individuals from different species or distinct populations, producing hybrid offspring. Hybridization can generate **heterosis** or hybrid vigor, where hybrids outperform parent populations in growth, survival, or other traits. Some hybrid combinations also produce predominantly male offspring, valuable in species where males grow faster than females. However, not all hybrid combinations show favorable characteristics; some exhibit reduced performance or reproductive abnormalities. Successful hybridization requires identifying favorable species combinations through systematic testing.

Common carp and crucian carp hybrids demonstrate favorable characteristics in some cases. Various carp species hybrids have been produced experimentally, with variable results depending on specific combinations. Tilapia hybridization has been extensively explored, producing some favorable combinations used commercially.

Monosex production through genetic or hormonal sex manipulation addresses situations where one sex has significantly superior culture characteristics. In Nile tilapia, males grow approximately 30 to 40 percent faster than females while females divert energy into egg production and exhibit earlier maturation. All-male tilapia populations improve growth uniformity and total production while eliminating unwanted reproduction in grow-out systems.

Sex reversal techniques administer male hormones during early larval development, masculinizing genetic females so they develop as functional males. Treated fry are then grown to market, producing phenotypically all-male populations though some individuals are genetically female. Alternatively, genetic approaches produce supermales with YY sex chromosomes rather than normal XY males. When YY supermales breed with normal females, all offspring are male. Maintenance of YY broodstock enables continuous production of all-male seed. Various technical challenges accompany these approaches, requiring careful implementation.

Inbreeding avoidance is critical in breeding programs, as inbreeding between close relatives causes **inbreeding depression**, reducing growth rates, survival, and fitness. Maintaining diverse broodstock from multiple family lines, periodically introducing new genetic material, and tracking pedigrees prevents inbreeding. Closed breeding populations limited in size require deliberate management preventing accumulation of inbreeding.

Broodstock exchange among facilities introduces genetic diversity and avoids inbreeding in isolated populations. Periodic introduction of broodstock from other sources refreshes genetic diversity. However, disease risk from introduced fish requires appropriate quarantine and health certification.

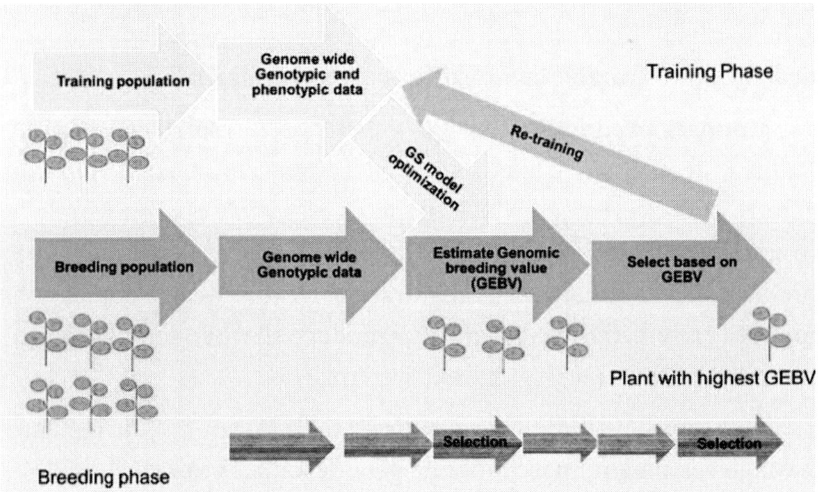

Figure. 5: Induced Breeding Process Sequence

2.5 Broodstock Management and Spawning Practices

High-quality broodstock maintained under optimal conditions form the foundation of successful breeding programs. Proper broodstock management encompasses selection, nutrition, health care, conditioning, handling, and maintenance practices ensuring broodfish reach excellent condition and produce high-quality gametes supporting successful reproduction.

Broodstock Selection Criteria

Selection of superior broodstock begins with identifying individuals exhibiting desirable characteristics that will be transmitted to offspring. Selection criteria vary somewhat among species and breeding objectives but generally emphasize characteristics including growth performance, body conformation, health status, maturity, and absence of deformities.

Growth performance represents a primary selection criterion, with rapidly-growing individuals favored as broodstock. Since growth rates are heritable, selecting fast-growing parents increases likelihood of offspring exhibiting superior growth. Individual growth records tracked from earlier rearing provide ideal information, though such detailed records are not always available. In their absence, selecting the largest individuals from a cohort of known age provides reasonable proxies for growth capacity.

Body conformation affects market quality and reproductive capacity. Ideal broodstock exhibit proper body form characteristic of the species, with good depth, appropriate length-to-depth ratios, and absence of skeletal deformities. Compressed, emaciated, or malformed individuals should be rejected regardless of size. Females with appropriate body depth and abdominal capacity can produce more eggs than shallow-bodied females of similar length.

Health status is critical, as diseased or parasitized broodstock produce poor-quality gametes, may transmit diseases to offspring, and perform unreliably in breeding. Broodstock should be examined carefully for external parasites, lesions, abnormal coloration, or behavioral indicators of disease. Only healthy, vigorous individuals merit selection.

Sexual maturity must be appropriate, with broodstock having reached maturity but not being excessively old. Optimal breeding age varies among species but generally includes individuals that have spawned successfully previously but are not senescent. First-time spawners sometimes produce lower egg quantities or quality compared to experienced breeders, though young, vigorous broodstock have advantages as well.

Genetic diversity should be considered when selecting broodstock to maintain genetic variation within cultured populations and avoid inbreeding depression. Selecting broodstock from multiple families or sources helps maintain diversity. When selective breeding programs are implemented, balancing selection intensity against maintenance of genetic variation requires careful management.

Sex identification enables selection of appropriate male and female broodstock. Sexual dimorphism—physical differences between sexes—varies among species. In carps, mature females typically exhibit rounder, softer abdomens particularly when containing developing eggs, while males may have more slender profiles. During breeding season, males of many species develop **breeding tubercles**—small raised bumps on head, fins, or body—and may exhibit brighter coloration. Genital papillae often differ between sexes, with females having larger, rounder, smoother genital openings while males have smaller, pointed papillae with a central slit. Cannulation or gentle abdominal pressure revealing eggs or milt provides definitive sex determination.

Broodstock Nutrition and Conditioning

Nutritional management profoundly affects broodstock condition, gonadal development, gamete quality, and spawning success. Broodstock require high-quality diets providing adequate energy, protein, essential fatty acids, vitamins, and minerals supporting maintenance metabolism, gonadal development, and accumulation of egg yolk reserves.

Protein requirements for broodstock generally equal or exceed those of growing fish, with requirements typically in the range of 35 to 45 percent dietary protein for most species. Adequate protein supports somatic maintenance while providing amino acids incorporated into developing eggs. High-quality protein sources with appropriate essential amino acid profiles are preferred.

Lipid nutrition is particularly important for broodstock, as lipids constitute major yolk components providing energy for embryonic development. Eggs contain substantial lipid reserves, requiring maternal diets supplying adequate lipids for yolk deposition. Essential fatty acids including omega-3 long-chain polyunsaturated fatty acids are particularly critical, affecting egg quality, fertilization success, and larval development. Broodstock diets should contain 8 to 15 percent lipid with appropriate ratios of omega-3 to omega-6 fatty acids.

Vitamin supplementation in broodstock diets exceeds requirements for grow-out fish. Fat-soluble vitamins including vitamins A, D, E, and K, and water-soluble vitamins including C and B-complex vitamins are required at elevated levels. Vitamin E particularly affects reproductive performance, with deficiency associated with poor gamete quality and reduced fertilization. Vitamin C supports egg quality and stress resistance. Commercial broodstock diets incorporate vitamin premixes formulated for reproductive requirements.

Mineral nutrition including calcium, phosphorus, zinc, and selenium affects reproductive function. Deficiencies in specific trace minerals can impair gonadal development and gamete quality. Most cultured fish obtain adequate minerals from water and feed under normal culture conditions, though supplementation may benefit broodstock.

Feeding management of broodstock maintains excellent condition throughout the year, with particular attention during gonadal development periods preceding breeding seasons. Daily feeding rates of 2 to 4 percent of body weight maintain condition in most species, with adjustments based on water temperature, fish condition, and seasonal reproductive status. Feeding multiple times daily improves feed utilization compared to single large meals. High-quality commercial broodstock feeds formulated specifically for reproductive requirements are preferred, though well-formulated grower feeds may suffice when specialized broodstock feeds are unavailable.

Body condition assessment monitors broodstock nutritional status, enabling feeding adjustments. Visually, well-conditioned broodstock appear robust with full, rounded bodies, bright coloration, and vigorous behavior. Emaciated fish with sunken bellies or prominent skeletal features indicate poor condition requiring nutritional intervention. **Condition factor** provides quantitative assessment, calculated from length and weight measurements. Values significantly below species-typical ranges indicate poor condition.

Broodstock Facilities and Environmental Management

Broodstock holding facilities should provide excellent environmental conditions, adequate space, minimal stress, and security against predators and theft. Facility design varies with available resources and operational scale, ranging from simple earthen ponds to sophisticated indoor recirculating systems.

Pond-based broodstock management is common in tropical and subtropical regions, utilizing earthen ponds dedicated to broodstock maintenance. Broodstock ponds should be of manageable size, typically 0.1 to 0.5 hectares, enabling good water quality management and facilitating broodfish capture when needed. Deeper ponds with maximum depths of 2 to 3 meters provide thermal stability and refuge from extreme temperatures. Water supply and drainage infrastructure enable water exchange and complete draining for fish harvest.

Stocking density in broodstock ponds should be conservative, maintaining fish comfort and excellent water quality. Densities of 500 to 2000 kilograms per hectare are typical for carp broodstock, well below grow-out densities. Lower densities reduce competition, stress, and water quality deterioration.

Water quality management for broodstock emphasizes maintaining optimal conditions. Dissolved oxygen should remain high, ideally near saturation, through aeration if necessary. Ammonia, nitrite, and other potentially toxic metabolites must remain at very low levels. pH and alkalinity should be appropriate for species requirements. Regular water exchange dilutes metabolites and replenishes oxygen. Monitoring water quality parameters and responding to problems promptly prevents stress and maintains broodstock condition.

Tank-based systems provide greater environmental control than ponds, enabling more intensive management and year-round breeding in species responsive to photoperiod or temperature manipulation. Fiberglass, concrete, or polyethylene tanks of various sizes can maintain broodstock. Flow-through or recirculating water systems maintain water quality. Environmental control systems regulate temperature, photoperiod, and water flow. Tank systems require greater investment and technical sophistication but offer advantages in controlled breeding programs.

Seasonal management accommodates changing environmental conditions and reproductive cycles. During non-breeding periods, broodstock maintenance focuses on nutrition, health, and condition improvement. As breeding season approaches, environmental conditions may be adjusted if breeding is seasonal. In species requiring temperature or photoperiod cues for maturation, gradual environmental changes can advance or delay breeding. Cool temperatures during winter may be beneficial for temperate species requiring cool periods for proper gamete maturation.

Pre-Spawning Preparation and Spawning Management

Preparation for spawning begins with assessment of broodstock maturity and readiness. Females are examined for evidence of ovarian development through external appearance and possibly cannulation or gentle abdominal pressure revealing mature eggs. Males are checked for milt production. Only fully mature broodstock exhibiting clear readiness should be selected for spawning attempts.

Broodstock selection and pairing for specific spawning attempts considers maturity status, size compatibility, genetic relationships, and breeding objectives. In selective breeding programs, specific males and females may be mated based on performance records and genetic evaluations. In commercial hatcheries without detailed genetic management, broodstock are selected based on maturity, condition, and size.

Handling broodstock requires careful technique preventing injury and minimizing stress. Fish are captured from ponds using seines or cast nets, gently restrained for examination, and transferred to transport containers or directly to breeding facilities. Overcrowding and oxygen depletion during handling cause severe stress and should be avoided. Anesthetic agents can be employed for extensive handling or examination, though use requires proper training.

Pre-spawning starvation for 24 to 48 hours before hormone injection is recommended practice, emptying the digestive tract and improving water quality in breeding containers. Starved fish release minimal fecal material into breeding enclosures, reducing contamination risk.

Hormone administration follows protocols appropriate for the species and induction method employed. As described previously, pituitary extracts, synthetic hormones, or commercial preparations are injected intramuscularly. Injection sites are swabbed with antiseptic before injection to reduce infection risk. Proper needle size and injection technique minimize tissue damage.

Breeding facilities receive injected broodstock after treatment. Breeding hapas suspended in ponds or breeding tanks with flowing water provide suitable spawning environments. Water circulation through breeding enclosures stimulates spawning behavior in many species. Lighting may be reduced to minimize stress and disturbance. Appropriate male-to-female ratios are established, typically two males per female for carps, though ratios vary.

Monitoring for spawning during the latency period following injection enables timely response. Broodstock are observed periodically without excessive disturbance. As spawning time approaches, increased activity, pursuing behavior, and close swimming indicate readiness. Spawning typically occurs during early morning hours, announced by increased splashing and activity.

Egg collection immediately following spawning prevents predation and damage. Fertilized eggs are gently collected from breeding containers using fine-mesh nets or by draining containers through collection nets. Eggs are examined for fertilization success, with fertilized eggs appearing translucent and swollen while unfertilized eggs remain opaque. Good fertilization rates should exceed 70 to 80 percent; substantially lower rates indicate problems with gamete quality, handling, or environmental conditions.

Broodstock recovery after spawning requires careful management supporting recuperation. Exhausted broodstock are returned to holding facilities with excellent water quality, minimal density, and reduced feeding initially. Careful monitoring for injury, infection, or stress-related disease enables early intervention if problems arise. Well-managed broodstock recover within several days and may be bred again after appropriate recovery periods, typically several weeks to months depending on species.

2.6 Seed Production and Hatchery Management

Efficient seed production through well-managed hatchery operations provides the foundation for expanded aquaculture production. Hatcheries bridge breeding programs and grow-out operations, transforming fertilized eggs into robust fry suitable for stocking. This section examines hatchery infrastructure, incubation systems, larval rearing practices, and management principles ensuring reliable seed supply.

Hatchery Infrastructure and Design

Hatchery facilities vary enormously in scale and sophistication from simple backyard operations producing thousands of fry to industrial facilities producing millions or tens of millions. Regardless of scale, effective hatcheries provide appropriate facilities for broodstock maintenance, spawning operations, egg incubation, larval rearing, and fry development.

Essential hatchery components include broodstock holding areas maintaining breeding stock, spawning facilities for breeding operations, incubation systems for egg hatching, larval rearing units for initial feeding and growth, and outdoor nursery facilities for fry development. Supporting infrastructure includes water supply and treatment systems, aeration equipment, electrical supply, feed storage, laboratory space for egg examination and water quality testing, and staff facilities.

Water supply is perhaps the most critical hatchery requirement, as reliable high-quality water in adequate quantities is essential for successful operations. Water sources include wells, springs, streams, or reservoirs. Water quality should meet stringent requirements including appropriate temperature ranges, high dissolved oxygen, freedom from pollutants, and appropriate chemistry. Flow-through systems require substantial continuous flow, while recirculating systems reduce water requirements through treatment and reuse but require more sophisticated infrastructure.

Water treatment may be necessary depending on source water quality. Aeration increases dissolved oxygen and strips carbon dioxide. Sedimentation removes suspended solids. Filtration through sand or other media removes finer particles. Disinfection using chlorine, ozone, or ultraviolet irradiation reduces pathogen risk, though chlorine must be removed through aeration or chemical neutralization before water contacts fish. Temperature control through heating or cooling may be necessary in some facilities and climates.

Hatchery building shelters sensitive incubation and early rearing stages from environmental extremes, predators, and contamination. Simple structures providing shade, rain protection, and wind protection suffice in many tropical locations. Fully enclosed, climate-controlled buildings enable year-round operation in temperate regions. Building design should facilitate efficient workflow from breeding through fry production, with logical spatial arrangement of functional areas.

Biosecurity measures prevent disease introduction and maintain healthy culture conditions. Biosecurity protocols include screening or filtering incoming water, disinfecting equipment and containers, limiting personnel access, requiring footbaths and hand washing, quarantining new broodstock, and maintaining separate water systems for different life stages. Rigorous biosecurity prevents catastrophic disease outbreaks that can destroy entire production cycles.

Egg Incubation Systems

Incubation systems provide appropriate environmental conditions for embryonic development from fertilization through hatching. Different incubation approaches suit different species based on egg characteristics including size, density, adhesiveness, and developmental requirements.

Trough or raceway incubation is common for non-adhesive, semi-buoyant eggs including carp eggs. Shallow troughs or tanks with water flowing lengthwise receive fertilized eggs that circulate in the gentle current. Water flow provides oxygen, removes metabolic wastes, and keeps eggs in gentle motion preventing fungal growth and maintaining uniform conditions. Flow rates are adjusted to maintain eggs in suspension without washing them downstream. The eggs tumble gently through incubation until hatching.

Hatching jars or **McDonald jars** provide vertical cylindrical containers where water enters at the bottom, flows upward through suspended eggs, and exits at the top. The upward flow keeps eggs in gentle circulation. This design works well for semi-buoyant eggs and is widely used for carp species. Jars of various sizes accommodate different volumes of eggs. Multiple jars supplied from common water systems enable simultaneous incubation of multiple spawns.

Hapa or net incubation employs fine-mesh cloth or net hapas suspended in ponds or tanks. Fertilized eggs placed in hapas develop in the pond or tank water environment while confined for easy monitoring and harvest. Gentle water circulation through hapas from pond or tank water movement provides oxygen and waste removal. This simple, low-cost method suits conditions where sophisticated incubation facilities are unavailable.

Tray incubation for adhesive eggs places eggs on trays or in containers where they remain stationary during development. Water flows over the eggs, providing oxygen and removing wastes. This approach suits catfish and other species with adhesive or demersal eggs. Trays may be stacked vertically in flow-through systems, increasing capacity within limited space.

Incubation density affects survival and development, with optimal densities varying among species and systems. Excessive density causes oxygen depletion, metabolite accumulation, and increased fungal or bacterial growth. Insufficient density underutilizes incubation capacity. Densities for carp eggs in jars commonly range from 200,000 to 500,000 eggs per jar of 10 to 20 liter capacity, depending on egg size and water flow.

Incubation temperature profoundly affects development rate and hatching success. Each species exhibits optimal temperature ranges supporting best survival and normal development. Maintaining stable temperatures within optimal ranges improves outcomes. Temperature monitoring and possible temperature control enable maintenance of proper conditions.

Water quality during incubation requires careful management. Dissolved oxygen should remain near saturation. Ammonia and carbon dioxide must be kept low through adequate water flow. Appropriate pH and alkalinity support development. Water hardness affects egg membrane permeability and osmoregulation. Free chlorine, heavy metals, and other contaminants must be absent. Regular water quality monitoring and system maintenance ensure proper conditions.

Fungal control addresses fungal infections affecting dead eggs and sometimes spreading to healthy eggs. Keeping eggs in gentle motion through water flow inhibits fungal growth. Removing dead eggs promptly prevents fungal spread. Antifungal treatments including malachite green, methylene blue, salt, or other approved agents can be applied, though use requires following proper dosages and regulations. Good water quality, appropriate densities, and gentle handling minimize egg mortality and associated fungal problems.

Hatching typically occurs 24 to 72 hours after fertilization depending on species and temperature. Emerging larvae break free from egg membranes and initially rest on the container bottom or drift in water currents. Newly hatched larvae, called **hatchlings** or **yolk-sac larvae**, possess attached yolk sacs providing nutrition during the first several days of life. After hatching, environmental requirements change somewhat, and management shifts to larval rearing.

Larval Rearing and Fry Development

Larval rearing encompasses the critical period from hatching through transition to exogenous feeding and development into fry. This period determines ultimate seed production success, as larval mortality can be substantial if conditions are improper.

Yolk-sac stage immediately follows hatching, during which larvae depend entirely on absorbed yolk providing energy and nutrients. During this period, larvae complete development of functional mouth parts, digestive system, and other structures necessary for independent feeding. Yolk-sac duration varies from two to five days depending on species and temperature. Environmental requirements during the yolk-sac stage emphasize clean water, appropriate temperature, adequate oxygen, and minimal disturbance. Feeding is unnecessary and actually harmful since larvae cannot yet ingest food.

First feeding represents a critical transition as yolk reserves deplete and larvae must successfully initiate exogenous feeding. Failure to feed promptly after yolk absorption leads to starvation and massive mortality. **First feed timing** requires observation of larval development. When yolk is nearly absorbed, the mouth has opened, and larvae begin actively swimming and exhibiting feeding behaviors, it is time to provide initial feed. For many species, this occurs three to five days post-hatching.

Live foods are essential first feeds for most fish larvae, as newly hatched larvae cannot digest artificial feeds. Larvae require small, highly nutritious, digestible food items matching their gape size and feeding abilities. **Zooplankton** including rotifers for tiny-mouthed larvae or larger cladocerans and copepods for species with larger mouths constitute classical first feeds. Many hatcheries produce zooplankton specifically for larval feeding through culture in separate ponds or tanks fertilized to produce phytoplankton supporting zooplankton populations.

Artemia nauplii—newly hatched brine shrimp larvae—provide widely used live feeds valued for consistent availability, appropriate size, high nutritional value, and ease of production. Artemia cysts are purchased dried, then hatched in aerated salt water within 24 to 48 hours, producing millions of nauplii per gram of cysts. Newly hatched nauplii are collected, rinsed, and fed to fish larvae. The broad use of Artemia in global aquaculture reflects its effectiveness as a larval feed.

Feeding frequency and density during early larval stages are critical. Larvae have small digestive capacities but high metabolic rates, requiring frequent feeding. Feeding three to six times daily maintains food availability. Feed density should ensure larvae can capture food items readily without requiring extensive searching but should not be so dense as to degrade water quality. Visual observation of larval feeding activity and gut fullness guides feeding decisions.

Transition to formulated feeds occurs as larvae grow larger and their digestive systems mature, enabling consumption and digestion of artificial feeds. **Micro-diets** or **starter feeds** formulated specifically for larval fish are introduced gradually, initially mixed with live feeds. The proportion of artificial feed increases over several days while live feed is reduced, completing the transition. Successful transition depends on proper timing relative to larval development, gradual changeover preventing abrupt dietary shifts, and high-quality starter feeds with appropriate particle sizes, digestibility, and nutritional composition.

Weaning refers to the complete transition from live to artificial feeds. Successfully weaned fry feeding entirely on formulated diets are ready for transfer to nursery facilities. Weaning typically occurs at ages of two to four weeks post-hatching depending on species and management. Weaned fry are substantially larger and more robust than newly hatched larvae, with improved survival likelihood in subsequent rearing.

Larval rearing systems include various approaches depending on species, hatchery capabilities, and scale. **Tank rearing** maintains larvae in tanks with flow-through or recirculating water, artificial feeding, and intensive management. This approach enables high densities and good control but requires sophisticated infrastructure. **Pond rearing** stocks larvae directly into ponds with abundant natural zooplankton, enabling extensive rearing with minimal intervention but less control over density and growth. **Hapa rearing** maintains larvae in fine-mesh enclosures suspended in ponds, combining partial confinement with access to natural productivity. Hybrid systems combine these approaches, perhaps starting with tank rearing followed by pond nursery.

Environmental management during larval rearing emphasizes maintaining excellent water quality, appropriate temperature, adequate oxygen, and minimal stress. Small larvae are extremely sensitive to environmental problems, with water quality deterioration causing rapid mortality. Low stocking densities, frequent water exchange or vigorous filtration, and careful feeding management maintain conditions. Temperature should remain stable within optimal ranges. Dissolved oxygen must stay high, ideally near saturation. Ammonia, nitrite, and other metabolites must remain extremely low.

Grading and culling during larval rearing removes deformed individuals and may separate size classes improving uniformity. Deformed larvae with spinal curvatures, missing fins, or other abnormalities are removed to prevent their inclusion in sold seed. Size grading reduces cannibalism in species where larger individuals prey on smaller siblings and improves growth uniformity by grouping similar-sized individuals.

Nursery Rearing and Fry Production

Nursery rearing develops weaned fry into larger juveniles suitable for stocking in grow-out facilities. This intermediate rearing period bridges hatchery and farm, producing robust seed with improved survival in final culture environments.

Nursery ponds typically are small earthen ponds ranging from 0.02 to 0.2 hectares, shallow with depths of 0.5 to 1.5 meters, and easily drained for harvest. Ponds receive weaned fry at densities appropriate for species and management intensity, typically several hundred thousand to several million per hectare. Fry are fed formulated feeds at high rates supporting rapid growth. Water quality is managed through aeration, fertilization, or water exchange as necessary.

Nursery duration commonly ranges from three to eight weeks depending on species growth rates and desired final size. During nursery rearing, fry grow from perhaps 10 to 20 millimeters at stocking to 30 to 80 millimeters at harvest. This increased size and robustness substantially improves survival when stocked into grow-out ponds or tanks where predation, competition, and environmental challenges are greater.

Feeding management in nursery operations emphasizes rapid growth through generous feeding of nutritionally complete starter or grower diets. Daily feeding rates of 5 to 15 percent of body weight are common, adjusted based on fish response, water temperature, and water quality. Multiple daily feedings improve feed utilization compared to single meals. Feed particle size is adjusted as fish grow, transitioning from fine crumbles to larger pellets matching fish mouth size.

Water quality management in nurseries requires attention similar to larval rearing but is somewhat less critical as juveniles are hardier than larvae. Adequate dissolved oxygen, low ammonia and nitrite, appropriate pH, and absence of contaminants support healthy growth. Aeration may be necessary, particularly at high stocking densities or during warm periods. Water exchange supplements aeration in flow-through systems.

Harvesting and handling seed from nurseries requires care to minimize stress and physical damage. Ponds are drained gradually, concentrating fish in harvest areas where they are collected using nets. Handling should be gentle, avoiding rough handling or prolonged air exposure. Harvested seed are often held in clean water without feed for 12 to 24 hours before transport, allowing digestive tract clearing reducing metabolic waste production during transport.

Packing and transport of seed from hatcheries to customers requires maintaining fish health during potentially long transit times. Oxygen-inflated plastic bags are standard transport containers, with appropriate fish densities, water volumes, and oxygen levels ensuring survival. Temperature control during transport prevents thermal stress. Transportation duration should be minimized. On arrival, gradual acclimation to receiving water temperature prevents shock.

Hatchery management and record-keeping enable continuous improvement and problem diagnosis. Records documenting breeding dates, broodstock identities, environmental conditions, fertilization rates, incubation survival, larval survival, feed consumption, growth rates, and final fry production provide valuable information evaluating performance and identifying problems. Analysis of records reveals patterns enabling management refinement.

Disease prevention through biosecurity, sanitation, and stress minimization is far more effective than attempting to treat disease outbreaks in hatcheries. Biosecurity protocols prevent pathogen introduction. Regular cleaning and disinfection of containers, nets, and facilities between production cycles eliminates disease carryover. Maintaining excellent water quality and minimizing stress through gentle handling and appropriate densities support robust larval immune function. If disease problems occur despite prevention efforts, isolating affected groups, diagnostic identification of pathogens, and appropriate treatments may mitigate losses.

Table 3: Critical Stages in Seed Production

Stage	Duration	Critical Factors	Primary Risks
Egg Incubation	1–3 days	Temperature, oxygen, clean water	Fungal infection, poor fertilization
Yolk-sac Larvae	2–5 days	Stable conditions, minimal disturbance	Premature feeding attempts, stress
First Feeding	3–7 days	Appropriate live feeds, frequent feeding	Starvation, inadequate feed density
Weaning	2–4 weeks	Gradual transition, quality micro-diets	Digestive problems, nutritional deficiency
Nursery Rearing	3–8 weeks	Optimal feeding, water quality	Predation, disease, poor growth
Harvest and Transport	Variable	Careful handling, oxygen supply	Physical injury, transport stress

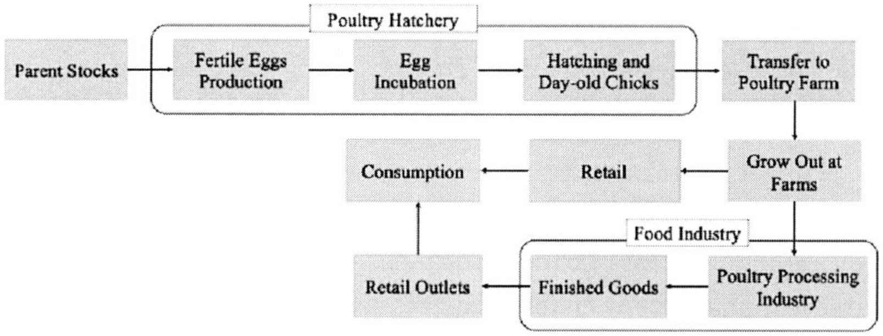

Figure.6: Hatchery Layout and Workflow

Conclusion

This comprehensive chapter has established essential biological foundations and practical methodologies for successful breeding and seed production in cultivable fishes. Understanding fish biology including anatomy, physiology, growth patterns, and particularly reproductive biology provides the knowledge base supporting effective breeding programs. The systematic approach to species selection considering biological characteristics, culture requirements, market potential, and environmental compatibility enables informed decisions matching species to operational contexts.

Reproductive biology of major cultivable species including Indian major carps, Chinese carps, tilapia, and catfishes reveals diverse strategies and requirements influencing breeding approaches. While some species breed naturally in captivity given appropriate conditions, major carps require induced breeding using hormonal preparations to achieve reliable reproduction. Mastery of both natural and induced breeding techniques, appropriate to different species and contexts, enables successful seed production under varied circumstances.

Broodstock management principles covering selection criteria, nutritional requirements, environmental management, and pre-spawning preparation ensure availability of high-quality breeding material capable of producing viable offspring. Proper broodstock care represents investment in seed production success, with neglected broodstock producing poor results regardless of subsequent hatchery expertise.

Hatchery operations transform fertilized eggs into robust fry suitable for stocking in grow-out operations. Understanding incubation requirements, larval rearing critical periods, feeding strategies, and nursery management enables reliable seed production. The integration of appropriate infrastructure, biological knowledge, and careful management creates efficient hatchery systems supporting aquaculture expansion.

The techniques and principles presented in this chapter provide practitioners with the knowledge necessary to implement successful breeding programs scaled appropriately to their resources and objectives. Whether operating simple backyard hatcheries or sophisticated industrial facilities, the biological understanding and technical competence developed through study of this material enables effective seed production supporting productive, profitable aquaculture operations. Subsequent chapters will build on this foundation, examining grow-out culture systems, nutrition and feeding, health management, and other aspects of integrated fish production systems.

CHAPTER 3

CULTURE SYSTEMS AND POND MANAGEMENT

Key Outcomes

1. Hydrological (water availability and quality
2. Edaphic (soil type and permeability)
3. Climatic (temperature, rainfall)
4. Biological (presence of predators, disease risk)
5. Socio-economic (accessibility, labor, market proximity)
6. Gently sloping land (1–3%) enables gravity-based water management.
7. Moderately undulating terrain supports embankment pond construction using excavated soil, lowering infrastructure costs.

3.1 Site Selection and Pond Construction

The foundation of successful pisciculture rests upon the judicious selection of site and scientifically planned construction of culture facilities. **Site selection** represents the most critical preliminary decision in establishing a fish farming enterprise, as it determines the biological productivity, operational efficiency, and economic viability of the entire operation. A well-selected site minimizes operational costs, maximizes production potential, and ensures long-term sustainability of the aquaculture venture.

Principles and Criteria for Site Selection

Site selection for fish culture is a multidimensional evaluation process that integrates topographical, hydrological, edaphic, climatic, biological, and socio-economic considerations. The ideal site must provide optimal conditions for fish growth and health while minimizing construction and operational expenditures.

Topographical Requirements

The topography of land fundamentally influences pond construction feasibility, water management efficiency, and overall development costs. **Gently sloping terrain** with a gradient of 1-3% is considered optimal for pond construction. Such slopes facilitate gravity-driven water flow, enabling efficient pond filling, drainage, and water exchange without excessive pumping requirements. Moderately undulating land allows for the creation of embankment ponds where excavated soil can be utilized to construct dykes, thereby reducing construction costs.

Flat lands with gradients less than 0.5% may require extensive earthwork to establish proper drainage systems, increasing initial capital investment. Conversely, steeply sloping areas with gradients exceeding 5% necessitate substantial excavation and high embankments, which dramatically escalate construction costs and create structural stability concerns. The site should possess sufficient elevation above local water bodies to prevent flooding during monsoon or high-water events, typically maintaining at least 1-2 meters clearance above the highest recorded flood level.

Land area requirements vary according to the scale and intensity of culture operations. For commercial pond aquaculture, a minimum contiguous area of 2-5 hectares is generally recommended to achieve economies of scale and operational efficiency. However, the site should accommodate future expansion and include adequate space for ancillary facilities such as feed storage structures, processing units, hatcheries, worker accommodation, vehicle parking, and access roads. A buffer zone of at least 10-15 meters around production ponds is advisable to prevent contamination and facilitate maintenance operations.

Water Availability and Quality

Water represents the most crucial resource in aquaculture, and its reliable availability in adequate quantity and acceptable quality is non-negotiable for site selection. The water source must provide sufficient volume throughout the culture period, typically calculated as:

$$V = (A \times D) + (A \times E \times t)$$

Where V = total water requirement (m^3), A = total pond area (m^2), D = average pond depth (m), E = evaporation and seepage rate (m/day), and t = culture period (days).

For tropical freshwater aquaculture, water requirements typically range from 8,000 to 15,000 m^3 per hectare for the initial filling, with additional water needed to compensate for evaporation losses of 4-8 mm/day and seepage losses of 10-25 mm/day, depending on soil type.

Potential water sources include rivers, streams, reservoirs, lakes, groundwater through tubewells or wells, and municipal or canal systems. Each source presents distinct advantages and limitations. **Surface water sources** like rivers and streams generally offer abundant supply and require minimal pumping, but may carry suspended sediments, pollutants, disease organisms, and competing or predatory species. Groundwater sources typically provide cleaner water with stable temperature and chemical composition, but may contain elevated dissolved minerals, low dissolved oxygen, and high carbon dioxide or hydrogen sulfide concentrations requiring pre-treatment.

Water quality parameters must be evaluated comprehensively before site selection. Critical parameters include temperature range (optimal 22-32°C for most tropical species), pH (6.5-8.5), dissolved oxygen (>5 mg/L at source), total alkalinity (50-200 mg/L as $CaCO_3$), total hardness (50-300 mg/L as $CaCO_3$), and absence of toxic substances including heavy metals, pesticides, and industrial effluents. Water sources contaminated with agricultural runoff, industrial discharge, or domestic sewage should be strictly avoided.

Soil Characteristics

Soil type profoundly influences pond construction feasibility, water retention capacity, and overall productivity. The ideal soil for pond construction possesses adequate **clay content** to minimize seepage, sufficient structural stability to support embankments, and appropriate fertility to sustain natural food production.

Clay and clay-loam soils containing 20-40% clay particles are considered optimal for pond construction. These soils exhibit excellent water-holding capacity with acceptable seepage rates below 10-15 mm/day. Sandy soils with clay content below 10% are unsuitable due to excessive seepage, requiring expensive pond lining or soil amendments. Highly organic or peaty soils are equally problematic, as they decompose over time, causing pond subsidence and water quality deterioration.

Soil pH significantly affects pond productivity and should range between 6.5 and 8.0. Acidic soils with pH below 6.0 may require liming to neutralize acidity and enhance nutrient availability. Highly alkaline soils with pH above 9.0 or sodic soils with high exchangeable sodium can create unfavorable water chemistry and should be avoided.

Soil testing should be conducted at multiple locations across the proposed site, sampling at depths of 0-30 cm and 30-60 cm. Laboratory analysis should determine particle size distribution, pH, organic matter content, cation exchange capacity, and presence of acid sulfate materials. **Acid sulfate soils** containing pyrite minerals are particularly problematic, as oxidation during pond construction can generate sulfuric acid, drastically lowering pH and releasing toxic aluminum and iron concentrations.

Climatic Considerations

Climatic factors directly influence fish growth rates, culture practices, and production economics. Temperature represents the most critical climatic variable, as it governs metabolic rates, feeding intensity, growth, and disease susceptibility in cultured species. Most tropical and subtropical aquaculture species exhibit optimal growth at temperatures between 25-32°C.

Sites experiencing prolonged winter periods with temperatures below 15°C may require heated culture systems or selection of cold-tolerant species, significantly increasing operational costs. Extremely high summer temperatures exceeding 35°C can cause thermal stress and require shading or increased water exchange. Annual temperature variation should be evaluated to determine the effective growing season and production cycles.

Rainfall patterns affect water availability, pond management, and disease occurrence. Sites in regions with annual rainfall above 1000 mm typically ensure adequate water supply, though excessive rainfall can cause pond overflow, fish escapement, and water quality deterioration. Conversely, arid regions with rainfall below 500 mm annually may face water scarcity, requiring reliable supplementary sources. Seasonal distribution of rainfall should be considered, as concentrated monsoon patterns necessitate robust drainage infrastructure.

Wind velocity and direction influence water circulation, oxygen distribution, and evaporation rates. Moderate winds of 5-15 km/h promote beneficial mixing and oxygenation, while strong winds above 40 km/h can cause wave action that damages embankments and increases turbidity. Windbreaks using trees or artificial barriers may be necessary in exposed locations.

Biological and Environmental Factors

The site should be assessed for presence of wild fish populations, predatory birds, and other aquatic animals that may compete with or prey upon cultured species. Proximity to wetlands, natural water bodies, or wildlife corridors may increase predation pressure from birds, snakes, and mammals. While complete isolation is impractical, sites should minimize exposure to such threats.

Environmental regulations and protected area designations must be thoroughly investigated. Sites within or adjacent to national parks, wildlife sanctuaries, mangrove forests, or environmentally sensitive areas may face legal restrictions on aquaculture development. Conversion of agricultural land, especially paddy fields, may require specific permissions from local authorities.

The site should not be located in areas prone to natural disasters such as earthquakes, tsunamis, cyclones, or landslides. Historical records should be consulted to assess risks, and construction designs should incorporate appropriate safety measures where risks are unavoidable.

Socio-Economic and Infrastructure Considerations

Proximity to markets determines the freshness of product delivery and transportation costs, which significantly impact profitability. Sites within 50-100 km of major urban markets or having access to efficient cold chain facilities are preferable. However, land costs generally increase near urban areas, requiring careful economic analysis.

Availability of quality fingerlings or seed stock is essential for sustained operations. Sites located within reasonable distance (100-200 km) of reputable hatcheries or seed production centers reduce transportation stress on juveniles and ensure timely stocking. Similarly, proximity to feed mills and suppliers minimizes feed procurement costs and ensures freshness.

Skilled labor availability influences operational efficiency and production success. Rural areas may offer lower labor costs but limited technical expertise, while peri-urban locations provide better access to trained workers. The site should be accessible year-round via all-weather roads to facilitate input supply, product marketing, and emergency response.

Electrical power availability enables operation of aerators, pumps, feeders, and processing equipment. Sites with reliable grid electricity are preferable, though renewable energy alternatives like solar or wind power are increasingly viable for remote locations. Communication infrastructure including mobile connectivity facilitates market information access, technical support, and business transactions.

Land ownership and legal status must be unambiguous, with clear title documents free from disputes. Leased lands should have long-term agreements (minimum 10-15 years) to justify capital investments. All necessary permits and licenses for aquaculture operations, water extraction, and construction should be obtainable without protracted bureaucratic delays.

Community acceptance and social harmony are often overlooked but critical factors. Local communities should view the project positively, and potential conflicts regarding water usage, employment, or environmental impacts should be addressed proactively through stakeholder consultations. Sites where aquaculture may compete with critical agricultural water needs or drinking water sources should be avoided to prevent social tensions.

Types of Culture Facilities

Fish culture facilities vary widely in design, construction, and management, ranging from simple earthen ponds to sophisticated recirculating aquaculture systems. The selection of appropriate culture facility depends on species requirements, production objectives, available resources, and environmental conditions.

Earthen Ponds

Earthen ponds represent the most common and economical culture facilities in freshwater aquaculture, comprising over 90% of global pond-based fish production. These are water-impounding structures created by excavation, embankment construction, or a combination of both. Earthen ponds offer several advantages including lower construction costs, natural food production through fertilization, large-scale production capacity, and ease of harvesting.

Earthen ponds are classified based on their construction method into three primary categories. **Excavated ponds** or dug-out ponds are created entirely by digging below ground level, with excavated soil used to create peripheral embankments. These are suitable for flat terrain with high water tables and minimal slope. **Embankment ponds** or dike ponds are constructed primarily above ground level using soil to build retaining walls, appropriate for gently sloping land where gravity drainage is feasible. **Semi-excavated ponds** combine both techniques, with partial excavation and embankment construction, representing the most common design that balances drainage efficiency with construction economy.

Based on water supply and drainage capabilities, earthen ponds are further categorized as **drainable ponds** and **non-drainable ponds**. Drainable ponds feature inlet and outlet structures with independent control, allowing complete water drainage for harvest, maintenance, and pond bottom treatment. These are strongly preferred for intensive culture as they facilitate better management. Non-drainable ponds lack drainage capability, requiring either pumping or complete fish removal while retaining water, suitable primarily for extensive culture or locations where gravity drainage is impossible.

The typical dimensions of earthen ponds vary with culture intensity and management capability. For semi-intensive commercial production, ponds of 0.2-2.0 hectares are optimal, providing sufficient scale while maintaining management efficiency. Larger ponds above 5 hectares become difficult to manage, particularly for feeding, disease monitoring, and harvesting. Smaller ponds below 0.1 hectare are suitable for intensive culture with mechanical aeration but face higher construction costs per unit area.

Pond depth is determined by species requirements, climate, and management practices. Most freshwater species thrive in ponds with **average depth** of 1.5-2.5 meters. Shallow ponds below 1.0 meter experience excessive temperature fluctuations, aquatic weed proliferation, and low carrying capacity. Excessively deep ponds above 3.5 meters develop thermal stratification, poor bottom conditions, and difficult management. In tropical regions prone to high evaporation, slightly deeper ponds of 2.0-2.5 meters are preferable. Cold-climate regions may employ deeper ponds of 2.5-3.0 meters to provide thermal refuge during winter.

Pond Components and Structures

A properly designed earthen pond incorporates several essential structural components that facilitate efficient management and ensure structural integrity.

Embankments or dykes are elevated earthen structures that retain pond water and must possess adequate strength, stability, and impermeability. The embankment crest width should be at least 2-3 meters to allow vehicle movement for feeding, harvesting, and maintenance operations. Wider crests of 4-5 meters are advantageous for large ponds where vehicular access is essential.

The height of embankments should exceed the maximum water depth by at least 0.5-0.75 meters to provide **freeboard** against wave action, rainfall overflow, and prevent fish escapement. For example, a pond with 2.0 meters water depth requires embankment height of 2.5-2.75 meters from pond bottom.

Embankment slopes determine structural stability and erosion resistance. The inner slope (water side) typically ranges from 1:1.5 to 1:2.5 (vertical:horizontal), with gentler slopes providing better stability and reducing erosion. Steeper slopes conserve land but may experience slumping, particularly in sandy soils. The outer slope is usually gentler, ranging from 1:2 to 1:3, to enhance stability and facilitate maintenance. The actual slope selection depends on soil type, with clayey soils permitting steeper slopes than sandy soils.

Embankments should be constructed in layers of 15-20 cm thickness, with each layer compacted to achieve 85-90% compaction density. Proper compaction eliminates air pockets, reduces permeability, and enhances structural strength. The core of the embankment should utilize clay-rich soil for impermeability, while outer portions may incorporate less clayey material. Vegetation should be removed, and topsoil stripped before embankment construction to prevent organic matter decomposition and subsequent settlement.

Inlet structures control water entry into ponds and should be positioned to facilitate uniform water distribution and prevent stagnant zones. The inlet pipe or canal should enter the pond at an elevation allowing gravity flow when possible, reducing pumping costs. Inlet structures must incorporate screening (mesh size 2-5 mm) to prevent entry of predatory fish, competitors, and disease vectors. A simple inlet design comprises a screened vertical pipe with an elbow joint directing water horizontally into the pond, dissipating energy and preventing bottom erosion.

Outlet structures enable controlled water discharge and complete drainage for harvest and maintenance. The most efficient outlet design is the **monk or standpipe system**, consisting of a vertical structure with removable boards allowing water level adjustment. The outlet should be positioned at the deepest point of the pond, typically at a corner or along one end, with the pond bottom sloping 0.2-0.5% toward the outlet for complete drainage.

The outlet structure should incorporate multiple components. A screened intake prevents fish escapement during water exchange, while the vertical standpipe with grooves allows insertion of multiple boards to control water level precisely. An overflow spillway at the design water level prevents uncontrolled overflow during heavy rainfall. The discharge pipe should be sufficiently large (typically 150-250 mm diameter for ponds of 0.5-2.0 hectares) to enable complete drainage within 24-48 hours, facilitating efficient harvest operations.

For small ponds or low-intensity culture, simpler outlet designs like **siphon systems** may be adequate. These comprise pipes installed through the embankment at the desired depth, controlled by removable stoppers or valves. However, siphon systems do not allow water level adjustment and complete drainage unless multiple outlets at different elevations are installed.

Pond Bottom Configuration

The pond bottom should be properly graded to facilitate complete drainage, prevent stagnant zones, and support benthic productivity. A uniform slope of 0.2-0.5% toward the outlet ensures water drainage without leaving residual pools. Completely flat bottoms are undesirable as they impede drainage and create anaerobic zones.

The pond bottom in drainable ponds should feature a **drainage canal or collection trench** extending from the outlet structure, typically 0.5-1.0 meter deeper and 1-2 meters wider than the general pond bottom. This channel collects all water during drainage and concentrates fish at the outlet, dramatically simplifying harvest operations. The drainage canal gradient should be steeper (0.5-1.0%) than the general pond bottom.

Pond bottoms should be left unlined in most earthen pond systems to allow natural benthic communities to develop. However, highly permeable bottoms may require treatment with bentonite clay, compaction, or application of synthetic liners. Removal of vegetation, stumps, roots, and debris is essential before pond filling, as decomposing organic matter depletes oxygen and releases toxic metabolites.

Concrete Ponds and Tanks

Concrete structures are employed where land cost is prohibitive, intensive culture is practiced, or species require controlled environments. Concrete ponds offer advantages including impermeability, durability, ease of cleaning and disinfection, precise water management, and suitability for recirculating systems. However, they require substantially higher capital investment, ranging from 5-15 times the cost of equivalent earthen ponds.

Concrete ponds are particularly suitable for hatchery operations, nursery rearing, broodstock maintenance, ornamental fish culture, and high-value species production. They are constructed as circular, rectangular, or raceway configurations, each offering specific advantages. **Circular tanks** provide excellent water circulation with minimal dead zones, uniform waste accumulation at the center for easy removal, and structural strength with minimal reinforcement. Diameters typically range from 3-20 meters with depths of 1.0-2.5 meters.

Rectangular concrete ponds facilitate easier construction, space-efficient arrangement, and convenient grading during harvest. However, they develop corner dead zones requiring attention during design. Typical dimensions are 10-50 meters length, 5-20 meters width, and 1.0-2.0 meters depth. The length to width ratio should not exceed 5:1 to prevent flow short-circuiting.

Raceway systems are elongated rectangular structures with continuous water flow, providing high flow-through rates suitable for salmonids and other species requiring cold, highly oxygenated water. Raceways typically measure 10-100 meters length, 2-5 meters width, and 0.8-1.5 meters depth, with length to width ratios of 10:1 to 30:1. Water velocities of 10-30 cm/second maintain self-cleaning and provide exercise for cultured fish. Raceways require large water volumes (5-10 complete exchanges per hour) and are primarily employed in regions with abundant cold spring water or at hatcheries.

Concrete structures must incorporate proper reinforcement, curing, and waterproofing. Concrete strength should be at least M20 grade (20 N/mm^2 compressive strength), with appropriate steel reinforcement for structural integrity. Newly constructed concrete requires thorough curing for 21-28 days and leaching of alkaline compounds by filling and draining several times before stocking fish. Application of waterproof coatings or epoxy paints further enhances durability and facilitates cleaning.

Cage Culture Systems

Cage culture represents an intensive production method where fish are confined in mesh or net enclosures suspended within larger water bodies such as lakes, reservoirs, rivers, or coastal waters. Cages utilize existing water resources without requiring land ownership or construction of impoundments, making them economically attractive where suitable water bodies are available.

Cage structures typically consist of a rigid frame (bamboo, wood, galvanized steel, or HDPE pipes) supporting net enclosures of appropriate mesh size. Cage sizes vary enormously, from small experimental cages of 1-10 m^3 to large commercial cages exceeding 1000 m^3. For freshwater species, typical commercial cages range from 20-100 m^3. Cage dimensions commonly measure 3-5 meters length, 3-5 meters width, and 2-4 meters depth, though deeper cages provide better conditions.

The mesh size selection depends on fish size, with smaller mesh for juveniles and progressively larger mesh as fish grow. Initial stocking uses 5-10 mm mesh, transitioning to 15-40 mm mesh for grow-out. **Netting materials** include nylon, polyethylene, and galvanized wire mesh, each with distinct durability and fouling characteristics.

Cage culture offers several advantages including high stocking density (50-200 kg/m^3), easy feeding and monitoring, simplified harvesting, and reduced land requirements. However, cages face challenges including complete dependence on external feed, vulnerability to water quality deterioration in host water bodies, predator attacks, theft, storm damage, and potential environmental impacts through waste accumulation and escaped fish.

Cage site selection within a water body requires careful assessment. Sites should provide adequate depth (minimum 1.5-2 times cage depth to ensure bottom clearance), good water exchange (current velocity 10-30 cm/second), dissolved oxygen above 5 mg/L, and protection from extreme wave action. Overcrowding of cages causes localized water quality deterioration, and appropriate spacing of 15-30 meters between cage rows is essential.

Pen Culture Systems

Pen culture employs large enclosures created by barrier nets in shallow water bodies, essentially partitioning lakes, lagoons, or back-waters for fish culture. Pens are larger than cages, typically ranging from 500-5000 m^2 surface area, and extend from surface to bottom, enclosing the entire water column.

Pen construction involves installing vertical stakes (bamboo, wood-en posts, or iron pipes) at 2-5 meter intervals along the perimeter and attaching barrier nets. The netting must be buried 20-30 cm into the bottom sediment and extend 30-50 cm above water surface to prevent fish escapement. Pen culture represents a semi-intensive system, often combining supplementary feeding with natural food production within the enclosure.

Pens are particularly suitable for culture of herbivorous species in nutrient-rich water bodies, as they exploit natural productivity while providing some control over fish confinement. However, pens offer less control over water quality than cages, are vulnerable to flooding and water level fluctuations, and face higher predation risk. Pen culture is practiced extensively in tropical Asian lagoons and floodplain lakes for species like Labeo rohita, Cirrhinus mrigala, and tilapias.

Recirculating Aquaculture Systems (RAS)

Recirculating aquaculture systems represent the most intensive and technologically sophisticated culture approach, recycling and treat-ing water continuously to minimize usage and maximize control. RAS facilities reuse 90-99% of system water, with daily make-up water of only 1-10% to replace evaporation and filter backwash losses.

A complete RAS incorporates multiple treatment components in-cluding mechanical filtration to remove suspended solids, biological filtration for ammonia and nitrite conversion, aeration or oxygenation to maintain dissolved oxygen, pH adjustment, ultraviolet sterilization or ozonation for pathogen control, and temperature regulation through heating or cooling.

The core of RAS is the **biofilter**, which houses nitrifying bacteria that convert toxic ammonia to nitrite and subsequently to less toxic nitrate through the nitrification process:

$NH_3 + O_2 \rightarrow NO_2\times + 3H\times + 2e\times$ (Nitrosomonas spp.)

$NO_2\times + \frac{1}{2}O_2 \rightarrow NO_3\times$ (Nitrobacter spp.)

Biofilters require sufficient surface area for bacterial colonization, adequate oxygen supply (4-6 mg O_2 per mg NH_3-N converted), and retention time for complete nitrification. Common biofilter designs include trickling filters, rotating biological contactors, fluidized sand beds, and moving bed biofilm reactors, each with distinct efficiency and space requirements.

RAS enables fish production independent of climate and water availability, provides biosecurity through environmental control, facilitates waste management, and allows urban or industrial facility integration. However, capital costs are extremely high (often exceeding \$500-1000 per m^3 culture volume), technical expertise requirements are substantial, system failures can cause rapid catastrophic losses, and operational costs for energy and labor are significant. RAS is economically viable primarily for high-value species (ornamental fish, sturgeon, groupers, marine species) or situations where water or land availability severely constrains conventional culture.

Pond Construction Procedures

Systematic construction following engineering principles ensures pond structural integrity, longevity, and functionality. Pond construction typically proceeds through sequential phases from site preparation through final testing before stocking.

Site Preparation and Layout

Following site selection, detailed surveying establishes precise pond layout, elevations, and drainage patterns. Professional surveying using total stations or GPS equipment ensures accurate implementation of designs. Pond perimeters are marked using stakes and strings, with embankment centerlines, inner and outer boundaries, and elevations clearly indicated.

Vegetation clearing removes all trees, shrubs, grasses, and organic debris from the construction area. Complete removal of stumps and root systems is essential, as decomposing organic matter creates settlement problems and water quality issues. In previously forested areas, removal and burning of organic matter should be complete, followed by topsoil stripping to 10-15 cm depth.

The topsoil contains high organic content and lacks structural stability, making it unsuitable for embankment core construction. Stripped topsoil should be stockpiled separately for later use in landscaping, aquatic vegetation cultivation, or agricultural purposes. Subsoil with higher clay content and lower organic matter provides superior embankment material.

Excavation and Earthwork

Excavation proceeds according to design depths, slopes, and contours. Excavators, bulldozers, or draglines are employed for large commercial ponds, while small ponds may be constructed using manual labor or smaller equipment. Excavated material should be tested for suitability as embankment fill, with clayey material preferred for embankment cores and less suitable material used for outer layers or landscaping.

Excavation depth should account for post-construction settlement, typically adding 5-10% extra depth to the design. The pond bottom is shaped to achieve designed gradients toward the drainage canal. Excessive excavation should be avoided, as refilling and recompaction are difficult and costly. Under-excavation is more easily corrected by additional removal.

Material from excavation should be transported efficiently to embankment locations, minimizing hauling distance. The pond-to-embankment material balance must be calculated to ensure sufficient excavated material is available for embankment construction. For embankment ponds where most soil comes from excavation, the volume of excavated material should equal or exceed the volume required for compacted embankments. The relationship can be expressed as:

$$V(\text{excavated}) \times SF \geq V(\text{embankment})$$

Where V represents volume and SF is the shrinkage factor accounting for compaction (typically 0.8-0.9, indicating 10-20% volume reduction upon compaction).

Embankment Construction

Embankment construction begins after excavation completion and employs proper compaction techniques to ensure stability and impermeability. The foundation area where embankments will be constructed should be cleared of vegetation, topsoil removed, and the surface scarified to enhance bonding between foundation soil and embankment fill.

Core trench construction is recommended for embankments on permeable foundations. A trench 0.5-1.0 meter deep and 1-2 meters wide is excavated along the embankment centerline and backfilled with compacted clay to create an impermeable barrier preventing seepage beneath the embankment.

Embankment construction proceeds in horizontal layers or lifts of 15-20 cm loose thickness. Each layer is compacted using vibratory rollers, sheep-foot rollers, or manual rammers to achieve 85-90% of maximum dry density determined by Proctor compaction tests. Moisture content during compaction should be maintained near optimum moisture content (typically 12-18% for clayey soils), as both excessively dry and saturated soils compact poorly.

Compaction effort should be greatest in the embankment core, where impermeability is critical. Outer zones require compaction primarily for slope stability and erosion resistance. The embankment core should contain soil with clay content of 25-40% for optimal impermeability, while outer zones may incorporate less clayey material.

Construction of embankments in lifts with proper compaction contrasts sharply with inferior methods where loose soil is simply dumped and spread. Properly compacted embankments exhibit density of 1.7-1.9 g/cm^3, while uncompacted fills achieve only 1.2-1.5 g/cm^3, resulting in excessive settlement, piping failures, and seepage.

Installation of Inlet and Outlet Structures

Inlet and outlet pipes should be installed during embankment construction, with pipes passing through the embankment core. Anti-seepage collars or cutoff collars must be attached to pipes at 2-3 meter intervals. These collars consist of concrete or steel plates extending 30-50 cm perpendicular to the pipe, forcing seepage water to follow a longer path and reducing flow along the pipe exterior.

Pipes should be laid on a firm foundation, bedded in concrete if passing through major embankments, and the area around pipes must be carefully compacted using hand tampers. Construction equipment should never operate directly over buried pipes until sufficient embankment height provides protection.

Outlet structures like monks are constructed using brick masonry or reinforced concrete, with floors and walls adequately thick to resist soil pressure and water forces. The outlet structure should be anchored into undisturbed soil or founded on compacted fill, and a cutoff wall should extend beneath the structure to prevent undermining.

Slope Protection and Erosion Control

Embankment slopes, particularly the inner slope exposed to wave action, require protection against erosion. Several methods provide slope stabilization and erosion control. **Grass turfing** establishes dense grass cover on slopes, with species like Cynodon dactylon (Bermuda grass) or Vetiveria zizanioides (vetiver grass) providing excellent root binding and erosion resistance. Grass establishment requires seeding or turfing immediately after embankment completion, before heavy rainfall can cause erosion.

Riprap protection involves placement of stones, rocks, or concrete blocks on slopes to dissipate wave energy and prevent erosion. Riprap is particularly important in large ponds exposed to strong winds and in zones of concentrated wave action. Stone sizes should be adequate to resist dislodgement, typically 10-30 cm diameter, placed in layers 20-40 cm thick over a filter fabric or gravel layer.

Vegetation barriers using shrubs or trees on the outer slope and crest provide windbreaks, prevent erosion, and enhance farm aesthetics. However, large trees should not be planted directly on embankments, as root growth can create seepage channels. Trees should be planted at least 5 meters from embankment toes.

Pond Testing and Commissioning

Before stocking fish, newly constructed ponds must undergo testing and conditioning. The pond should be filled gradually, monitoring for leaks, embankment stability, and proper functioning of inlet and outlet structures. The first filling should proceed slowly over 3-5 days, allowing embankment settlement and identifying seepage problems while repair is still feasible.

Newly excavated ponds often exhibit poor water quality due to oxidation of reduced compounds in bottom soils. Excess turbidity, low pH, and elevated iron or aluminum may occur. Liming with agricultural limestone at 200-500 kg/ha and fertilization to establish algal blooms help condition new ponds. After filling, water should be retained for 7-10 days, then drained completely to allow the pond bottom to dry and oxidize. This cycle may be repeated 2-3 times before final filling for fish stocking.

Ponds constructed on acid sulfate soils require extensive leaching and liming. Multiple filling and draining cycles combined with heavy lime application of 2-10 tons/ha may be necessary to neutralize acidity and oxidize pyrite materials. Water pH should be stable above 6.5 before considering fish introduction.

All embankments should be inspected after the first filling for slumping, cracking, or seepage. Minor seepage through the embankment body often decreases as fine particles clog voids, but continuous seepage flows, piping failures where water emerges in concentrated streams, or progressive erosion require immediate repair. Affected sections must be drained, excavated, and properly reconstructed with improved compaction.

The outlet structure functionality should be verified by adjusting water levels and confirming smooth operation of boards or gates. Complete drainage should be tested to ensure the pond empties within the design timeframe and that the drainage channel effectively concentrates water and fish at the outlet.

Applied Implications for Pond-Based Aquaculture

The selection of appropriate sites and execution of proper pond construction determine the biological potential and economic viability of fish culture operations. Well-designed and properly constructed ponds minimize operational problems, reduce maintenance costs, enhance production efficiency, and support sustainable intensification. Investment in professional surveying, quality construction materials, proper compaction equipment, and attention to engineering details during initial construction provides long-term benefits far exceeding the incremental costs. The biological productivity and economic returns achievable in scientifically planned and constructed culture facilities consistently exceed those from inferior facilities, validating the critical importance of applying sound engineering principles to pisciculture infrastructure development.

3.2 Water Quality Parameters and Management

Water serves as the life-supporting medium in aquaculture, governing every physiological process from respiration and osmoregulation to feeding, growth, and reproduction. **Water quality** encompasses the physical, chemical, and biological characteristics that determine the suitability of water for supporting aquatic life. Understanding and managing water quality parameters represents perhaps the most critical technical skill in pisciculture, as suboptimal conditions directly impair fish health, growth efficiency, and survival, ultimately determining production success or failure.

Physical Parameters of Water Quality

Physical properties of water including temperature, transparency, turbidity, and color profoundly influence fish physiology, pond ecology, and management practices.

Temperature

Water temperature represents the most influential physical parameter, regulating metabolic rates, feeding intensity, growth, immune function, reproductive maturation, and disease susceptibility in cultured fishes. Fish are **poikilothermic** organisms whose body temperature approximates ambient water temperature, making them exceptionally sensitive to thermal conditions.

Each species possesses a characteristic temperature tolerance range bounded by lower and upper lethal limits beyond which survival becomes impossible. Within the tolerance range, an **optimum temperature range exists where physiological processes operate most efficiently, typically spanning 8-12°C for cold-water species like trout and salmon, 18-24°C for cool-water species like common carp, and 26-32°C for warm-water tropical species including tilapia, catfish, and major Indian carps.

Temperature directly influences metabolic rate through the temperature coefficient or Q_{10} value, which expresses the rate change for each 10°C temperature increase. For most fish species, Q_{10} values range from 2.0 to 2.5, indicating that metabolic rates approximately double with each 10°C temperature rise within the optimal range. This relationship can be expressed as:

$$R_2 = R_1 \times Q_{10}\wedge((T_2 - T_1)/10)$$

Where R_1 and R_2 represent metabolic rates at temperatures T_1 and T_2 respectively. This exponential relationship explains why feeding rates, growth rates, and oxygen consumption increase dramatically with temperature elevation within the optimal range.

Temperature affects dissolved oxygen solubility inversely, with cold water holding substantially more oxygen than warm water. At atmospheric pressure, oxygen solubility decreases from approximately 14.6 mg/L at 0°C to 9.1 mg/L at 20°C and 7.6 mg/L at 30°C. This inverse relationship creates a physiological paradox where fish at elevated temperatures exhibit increased oxygen demand due to accelerated metabolism but face reduced oxygen availability.

Sudden temperature fluctuations exceeding 3-5°C can induce thermal shock, causing stress, immunosuppression, and mortality. Temperature changes should be minimized during fish transport, grading, and water exchange operations. Acclimation procedures equalizing temperatures gradually over 15-30 minutes prevent thermal shock.

Thermal stratification develops in deep ponds and lakes during warm periods when surface water heated by solar radiation becomes less dense and forms a distinct warm layer above cooler bottom water. The **thermocline** represents a transition zone of rapid temperature decline separating the warm upper layer (epilimnion) from the cold bottom layer (hypolimnion). Stratified water bodies exhibit poor vertical mixing, leading to oxygen depletion in bottom layers as organic matter decomposes without replenishment from atmospheric sources. Destratification through mechanical aeration or circulation becomes necessary to prevent anoxic conditions.

Management of temperature in aquaculture systems involves several strategies. Pond depth influences temperature stability, with deeper ponds of 2.0-2.5 meters exhibiting less diurnal fluctuation than shallow ponds. Shading using nets, floating vegetation, or peripheral trees reduces summer temperature extremes. Water exchange during midday brings cooler source water into ponds, though this must be balanced against water availability and pumping costs. In cold climates, greenhouse covering or heated water systems maintain adequate temperatures for year-round production, though energy costs can be prohibitive.

Transparency and Turbidity

Water transparency, measured using a **Secchi disk**, indicates the depth of light penetration and relates inversely to turbidity. The Secchi disk consists of a white disk 20-30 cm in diameter lowered into water until it disappears from view, with the depth of disappearance recorded as Secchi depth. Optimal Secchi depth for fish ponds ranges from 25-45 cm, indicating moderate planktonic productivity without excessive turbidity.

Shallow Secchi depth below 20 cm indicates high turbidity from either planktonic blooms or suspended clay particles. While moderate phytoplankton blooms provide natural food and oxygen production, excessive density creates instability with risk of sudden die-offs causing oxygen depletion. Very clear water with Secchi depth exceeding 60 cm suggests low productivity requiring fertilization to stimulate natural food production.

Turbidity refers to water cloudiness caused by suspended particles. **Biological turbidity** from planktonic organisms is generally beneficial at moderate levels, contributing to pond productivity. **Clay turbidity** from suspended inorganic particles reduces light penetration, inhibits photosynthesis, clogs fish gills, and provides no nutritional value. Clay turbidity commonly occurs in newly constructed ponds, after heavy rainfall, or in ponds with unstable banks. Management of clay turbidity includes pond bottom stabilization, grassing of banks, reduced water exchange during turbid periods, and application of gypsum or alum to flocculate suspended particles.

Gypsum ($CaSO_4 \cdot 2H_2O$) application at 200-500 kg/ha causes clay particles to aggregate and settle, clearing turbid water within 2-3 days. Aluminum sulfate or alum ($Al_2(SO_4)_3$) at 15-25 mg/L similarly promotes flocculation but must be used cautiously as excess aluminum can be toxic to fish.

Color

Natural water bodies exhibit various colors depending on dissolved substances and suspended materials. Greenish coloration indicates healthy phytoplankton populations, particularly green algae and diatoms, suggesting productive conditions. Golden-brown water suggests dominance of diatoms and dinoflagellates. Blue-green coloration often indicates cyanobacterial blooms, which may produce toxins harmful to fish and pose management challenges.

Brown or tea-colored water results from dissolved humic substances and organic acids, common in ponds surrounded by forests or having high organic inputs. While moderate coloration is acceptable, dark brown water limits light penetration and productivity. Red or reddish-brown coloration may indicate iron oxidation or blooms of certain algae, requiring investigation of underlying causes.

Sudden color changes, particularly to milky white or grayish, may signal bacterial blooms associated with organic overload and deteriorating water quality. Regular monitoring of color changes provides early warning of developing water quality problems.

Chemical Parameters of Water Quality

Chemical characteristics fundamentally determine water's capacity to support fish life, with several parameters requiring careful monitoring and management.

Dissolved Oxygen (DO)

Dissolved oxygen represents the most critical and dynamically variable water quality parameter. Fish require oxygen for aerobic respiration, with inadequate levels causing stress, reduced feeding and growth, disease susceptibility, and mortality. Most cultured species require minimum DO concentrations above 5 mg/L for optimal growth, though survival is possible at reduced levels.

Oxygen enters water through two primary mechanisms: atmospheric diffusion across the air-water interface and photosynthetic production by phytoplankton and aquatic plants. Atmospheric diffusion occurs slowly and is driven by the oxygen deficit between saturation level and actual concentration. The diffusion rate can be approximated by:

$$dC/dt = K_L \times (C_s - C_t)$$

Where dC/dt represents the oxygen transfer rate, K_L is the liquid film mass transfer coefficient, C_s is saturation concentration, and C_t is actual concentration at time t. In stagnant water, atmospheric diffusion contributes only 0.2-0.5 mg/L per hour, insufficient to meet demand in productive ponds. Wind-driven surface turbulence and mechanical aeration substantially increase K_L and oxygen transfer rates.

Photosynthetic oxygen production by phytoplankton dominates oxygen dynamics in fertilized ponds. During daylight hours, photosynthesis may produce 5-15 mg/L or more oxygen, supersaturating surface waters to 150-200% of saturation. The photosynthetic oxygen production can be related to primary productivity:

$$O_2 \text{ produced (mg)} = \text{Net Primary Production (mg C)} \times (32/12) \times PQ$$

Where 32/12 represents the molecular weight ratio of oxygen to carbon, and PQ is the photosynthetic quotient (typically 1.1-1.3).

Oxygen consumption in ponds results from fish respiration, bacterial decomposition of organic matter, chemical oxidation, and night-time respiration of phytoplankton. Fish oxygen consumption increases with body weight, temperature, feeding rate, and activity level. Routine oxygen consumption for typical warm-water species approximates:

O_2 consumption (mg/kg/hr) = a × W^b × Q_{10}^((T-20)/10)

Where W is body weight in kg, a and b are species-specific constants (typically a = 200-400, b = -0.2 to -0.3), and Q_{10} is approximately 2.0-2.3.

Diurnal oxygen fluctuation in productive ponds exhibits characteristic patterns, with concentrations rising during afternoon to peak at 8-15 mg/L by late afternoon, then declining through evening and night to reach minimum levels of 2-4 mg/L or lower by dawn. This pattern reflects the dominance of photosynthetic production during daylight and continuous respiratory consumption throughout the 24-hour cycle.

Critical oxygen depletion occurs during predawn hours, particularly following cloudy days with reduced photosynthesis, in heavily stocked ponds, during algal bloom die-offs, or after application of excessive organic matter. **Oxygen crashes** represent catastrophic depletion events causing massive fish mortality if not promptly addressed.

Management of dissolved oxygen involves multiple strategies. Stocking density must be matched to the pond's oxygen-producing capacity, typically limiting biomass to 3,000-5,000 kg/ha in non-aerated ponds. Feeding rates should be adjusted based on oxygen availability, reducing or eliminating feeding when DO falls below 4-5 mg/L. Excessive organic inputs from feed wastage or manure must be avoided to prevent oxygen-consuming decomposition.

Mechanical aeration using paddlewheel aerators, propeller-aspirator-pump aerators, or diffused air systems provides supplemental oxygen and enables intensification beyond natural carrying capacity.

Aerator operation typically begins when DO falls to 3-4 mg/L, usually during late night or predawn hours. Aerator horsepower requirements depend on biomass, typically 2-3 HP per ton of fish in intensive culture. Emergency aeration must be available to respond to oxygen crises, as delays of even 1-2 hours can result in significant mortality.

Water exchange can alleviate oxygen deficiency if source water contains higher DO than pond water. However, source water from anoxic bottom layers of reservoirs or from groundwater may actually contain less oxygen than pond surface water, exacerbating rather than ameliorating the problem. Daytime water exchange may be preferable, as source water is likely to contain reasonable oxygen levels.

Reduction of organic loading through decreased feeding, removal of accumulated bottom organic matter, and avoidance of excessive fertilization prevents oxygen demand from exceeding supply capacity. Maintaining moderate rather than excessive plankton blooms (Secchi depth 30-40 cm) provides oxygen production while avoiding instability associated with dense blooms.

pH (Hydrogen Ion Concentration)

pH expresses water acidity or alkalinity on a logarithmic scale, defined as the negative logarithm of hydrogen ion concentration:

$$pH = -\log_{10}[H^+]$$

The pH scale ranges from 0 (extremely acidic) to 14 (extremely alkaline), with pH 7.0 representing neutrality. Most freshwater fish thrive in pH range of 6.5-8.5, with optimal growth typically occurring at pH 7.0-8.0.

pH influences multiple physiological and chemical processes in aquaculture systems. **Ammonia toxicity** increases dramatically with rising pH, as the equilibrium between toxic un-ionized ammonia (NH_3) and relatively harmless ammonium ion (NH_4^+) shifts toward the toxic form at alkaline pH. The fraction of total ammonia in the un-ionized form can be calculated as:

$$f(NH_3) = 1 / (1 + 10 \wedge (pK\times - pH))$$

Where pK× for the ammonia-ammonium equilibrium is approximately 9.25 at 25°C. At pH 7.0, only about 0.4% of total ammonia exists as toxic NH_3, but this increases to 4% at pH 8.0 and 25% at pH 9.0, elevating toxicity correspondingly.

Fish experience physiological stress at pH below 5.5 or above 9.5, with exposure to pH below 4.5 or above 10.5 causing acute toxicity. Low pH impairs blood oxygen transport, increases susceptibility to bacterial infections, and causes direct gill damage. High pH also damages gill tissues and disrupts osmoregulation.

pH exhibits diurnal fluctuation in productive ponds due to photosynthetic carbon dioxide uptake during daylight and respiratory carbon dioxide release during darkness. Morning pH readings may be 7.0-7.5, rising to 8.5-9.0 by late afternoon in heavily fertilized ponds. Fluctuations exceeding 1.5 pH units indicate unstable conditions requiring management intervention.

Carbon dioxide and carbonate chemistry fundamentally govern pH dynamics in aquaculture systems. Dissolved carbon dioxide reacts with water forming carbonic acid:

$$CO_2 + H_2O \times H_2CO_3 \times H\times + HCO_3\times \times 2H\times + CO_3^{2}\times$$

During photosynthesis, phytoplankton consume CO_2, shifting equilibria leftward and increasing pH. Conversely, respiratory CO_2 release shifts equilibria rightward, decreasing pH. The total alkalinity (carbonate-bicarbonate buffering capacity) determines pH stability during these diurnal fluctuations.

Management of pH problems depends on the underlying cause. **Acid ponds** with pH below 6.5 require liming to neutralize acidity and establish carbonate buffering. Agricultural limestone ($CaCO_3$) or quicklime (CaO) application raises pH and increases alkalinity. Liming rates depend on soil and water chemistry, typically ranging from 250-2,000 kg/ha. The lime requirement can be estimated from buffer pH tests conducted on pond bottom soils.

Alkaline ponds with pH above 9.0 indicate excessive planktonic photosynthesis or high carbonate/bicarbonate concentrations. Reduction of plankton density through decreased fertilization or herbivorous fish stocking moderates pH elevation. Application of organic acids, alum, or gypsum can temporarily reduce pH, though addressing underlying productivity is more sustainable.

Ponds constructed on acid sulfate soils face severe pH problems from sulfuric acid generation during pyrite oxidation. Multiple applications of heavy lime doses (5-20 tons/ha) combined with repeated flushing may be required, making such sites generally unsuitable for aquaculture unless extensive amelioration is feasible.

Alkalinity and Hardness

Total alkalinity measures water's capacity to neutralize acids, reflecting concentrations of carbonates, bicarbonates, and hydroxides, expressed as mg/L of calcium carbonate equivalent ($CaCO_3$). Alkalinity provides **pH buffering**, resisting pH changes when acids or bases are added. Waters with high alkalinity (>50 mg/L) exhibit stable pH, while low alkalinity waters (<20 mg/L) experience dramatic pH fluctuations.

Optimal alkalinity for freshwater aquaculture ranges from 50-200 mg/L as $CaCO_3$. Low alkalinity limits primary productivity and creates unstable pH, while extremely high alkalinity (>300 mg/L) may be associated with elevated pH problems. Total alkalinity can be measured through titration with standard acid to pH 4.5 endpoint.

Total hardness represents the concentration of divalent cations, primarily calcium and magnesium, also expressed as mg/L $CaCO_3$ equivalent. Hardness classification ranges from soft water (<50 mg/L), moderately hard (50-100 mg/L), hard (100-200 mg/L), to very hard (>200 mg/L). Most fish species grow well in moderately hard to hard water.

Calcium and magnesium serve essential physiological functions, contributing to bone and scale formation, blood clotting, enzyme activation, and osmoregulation. Deficiency in very soft water can impair growth and increase disease susceptibility. However, extremely hard water (>400 mg/L) may reduce egg hatching success in some species.

Alkalinity and hardness management primarily involves liming of ponds with low values. Agricultural limestone provides both calcium and carbonate, simultaneously increasing hardness and alkalinity. For waters where only hardness is low but alkalinity is adequate, gypsum (calcium sulfate) application increases calcium without substantially affecting alkalinity or pH.

Ammonia, Nitrite, and Nitrate

Nitrogenous compounds represent major metabolic waste products in aquaculture systems, with toxicity varying significantly among chemical forms. Fish excrete nitrogen primarily as ammonia through gill membranes, with production rates depending on protein intake, feed quality, and metabolic rate. Ammonia excretion approximates 20-30 mg of ammonia-nitrogen per kilogram fish per day, increasing with protein content of diets.

As previously discussed, **ammonia** exists in equilibrium between toxic un-ionized form (NH_3) and relatively harmless ionized ammonium (NH_4^+), with pH and temperature governing the equilibrium. Chronic exposure to un-ionized ammonia concentrations as low as 0.02-0.05 mg/L impairs growth and immune function, while acute toxicity occurs at levels exceeding 0.2-0.5 mg/L depending on species.

Ammonia toxicity manifests through multiple mechanisms including gill damage, disruption of blood pH regulation, neurotoxicity, and impairment of oxygen transport. Affected fish exhibit lethargy, loss of equilibrium, increased surface respiration, and elevated susceptibility to diseases.

In aerobic aquatic systems, ammonia undergoes **nitrification**, a two-step bacterial process converting ammonia to nitrite and subsequently nitrate:

$NH_4^+ + 1.5O_2 \rightarrow NO_2^- + 2H^+ + H_2O$ (Nitrosomonas)

$NO_2^- + 0.5O_2 \rightarrow NO_3^-$ (Nitrobacter)

Nitrite represents an intermediate oxidation product exhibiting moderate toxicity. Nitrite absorbed through gills enters the bloodstream and oxidizes hemoglobin to methemoglobin, which cannot transport oxygen, causing "brown blood disease." Chronic nitrite exposure at levels above 0.1-0.5 mg/L impairs growth, while acute toxicity occurs above 5-10 mg/L in freshwater species. Marine and brackish water species exhibit greater nitrite tolerance due to chloride's competitive inhibition of nitrite uptake.

Nitrate is the final nitrification product and exhibits minimal toxicity at concentrations below 100-200 mg/L. Nitrate accumulates in intensive recirculating systems where complete nitrification occurs without regular water exchange. Very high nitrate levels (>300 mg/L) may impair immune function and growth, necessitating partial water exchange or denitrification processes.

Management of nitrogenous wastes employs multiple approaches. **Optimizing feeding practices** reduces excess nitrogen inputs by matching feed quantities to fish requirements, using high-quality feeds with digestible proteins, and avoiding overfeeding. Feed conversion ratios below 1.5:1 indicate efficient protein utilization with minimal waste generation.

Biological filtration in intensive systems and recirculating aquaculture harnesses nitrifying bacteria to convert ammonia to nitrate. Biofilters require adequate surface area for bacterial colonization (typically 0.3-1.0 m^2 biofilter area per kg feed per day), sufficient dissolved oxygen (>4 mg/L), appropriate pH (7.0-8.0), and moderate temperatures (20-30°C).

Water exchange dilutes accumulated nitrogenous wastes in flow-through and semi-intensive systems. Daily water exchange rates of 5-20% typically maintain ammonia and nitrite at acceptable levels, though higher exchange rates may be necessary in very intensive culture. The required water exchange rate can be estimated from:

Q = (P × E) / (C_in - C_out)

Where Q is required flow rate, P is daily waste production, E is excretion rate, C_in is input concentration, and C_out is desired output concentration.

Phytoplankton assimilation in fertilized ponds removes significant ammonia as algae incorporate nitrogen for protein synthesis. Moderate plankton blooms can process 10-30 kg nitrogen per hectare per day, providing natural biological treatment. However, plankton communities are unstable, and sudden die-offs release accumulated nitrogen back into water.

Zeolite application provides emergency ammonia removal in crisis situations. Natural zeolites exhibit cation exchange capacity, adsorbing ammonium ions from water. Application rates of 100-500 kg/ha provide temporary ammonia reduction, though zeolites saturate and require replacement or regeneration.

Chloride supplementation at ratios of 10-20 mg $Cl\times$ per 1 mg $NO_2\times$-N reduces nitrite toxicity through competitive inhibition. Addition of salt (NaCl) to achieve chloride concentrations of 50-100 mg/L provides substantial protection against nitrite toxicity in freshwater systems.

Carbon Dioxide

Carbon dioxide (CO_2) represents the primary respiratory waste product and, when accumulated to excess concentrations, induces toxicity in fish. While carbon dioxide is less acutely toxic than ammonia, chronic exposure to elevated levels impairs growth, reduces feeding, and stresses fish.

Carbon dioxide in water exists in multiple forms based on pH:

$CO_2 + H_2O \times H_2CO_3$ (carbonic acid) $H_2CO_3 \times H\times + HCO_3\times$ (bicarbonate) $HCO_3\times\times H\times + CO_3{}^2\times$ (carbonate)

Free carbon dioxide (CO_2 and H_2CO_3) represents the fraction directly toxic to fish. Safe levels of free CO_2 remain below 10-15 mg/L for most species, though toxicity depends on oxygen availability. The ratio of oxygen to carbon dioxide should ideally exceed 5:1, as high CO_2 impairs oxygen uptake and transport even when oxygen levels appear adequate.

Carbon dioxide toxicity manifests as respiratory acidosis, where blood pH decreases due to CO_2 accumulation, impairing metabolic functions. Affected fish exhibit increased ventilation rates, lethargy, and reduced growth. Chronic exposure to 15-25 mg/L CO_2 reduces growth rates by 10-30% even with adequate dissolved oxygen.

Carbon dioxide accumulation occurs in densely stocked ponds with poor circulation, in deep stratified ponds where respiration exceeds photosynthesis in bottom layers, and in intensive indoor systems with inadequate aeration. Groundwater sources often contain elevated CO_2 (20-50 mg/L) from subterranean respiration and require aeration before use in fish culture.

Management of carbon dioxide involves vigorous aeration or agitation that strips excess CO_2 from water. Cascade aerators, fountain aerators, and vigorous surface agitation promote gas exchange, allowing CO_2 to escape to atmosphere while simultaneously increasing dissolved oxygen. Maintaining adequate phytoplankton populations in ponds provides CO_2 consumption through photosynthesis. Reduced stocking densities and feeding rates decrease respiratory CO_2 production. Water exchange with low-CO_2 source water dilutes accumulated carbon dioxide.

Hydrogen Sulfide

Hydrogen sulfide (H_2S) is an extremely toxic metabolic byproduct formed under anaerobic conditions through bacterial reduction of sulfate compounds. Hydrogen sulfide generation occurs in pond bottom sediments where organic matter accumulation creates oxygen-depleted zones supporting sulfate-reducing bacteria:

$SO_4^{2}\times$ + Organic matter → $H_2S + CO_2 + H_2O$

Similar to ammonia, hydrogen sulfide exists in pH-dependent equilibrium between toxic un-ionized H_2S and less harmful bisulfide ($HS\times$) and sulfide ($S^{2}\times$) ions. Acid conditions favor toxic H_2S formation, while alkaline conditions shift equilibrium toward ionic forms.

Un-ionized hydrogen sulfide is extraordinarily toxic, causing acute mortality at concentrations as low as 0.01-0.05 mg/L. Hydrogen sulfide toxicity results from inhibition of cytochrome oxidase enzymes essential for cellular respiration, essentially causing cellular asphyxiation despite adequate environmental oxygen.

Hydrogen sulfide problems typically occur in ponds with heavy organic loading, poorly oxidized bottom sediments, stratified waters where bottom layers become anoxic, and during pond turnover events when anoxic bottom water suddenly mixes with surface layers. The characteristic rotten egg odor signals hydrogen sulfide presence.

Management focuses on preventing anaerobic conditions. Adequate dissolved oxygen maintenance through aeration prevents sulfate reduction. Periodic draining and drying of pond bottoms oxidizes accumulated organic matter and converts sulfide compounds to non-toxic sulfate. Application of oxidizing agents like potassium permanganate (2-4 mg/L) or hydrogen peroxide (5-10 mg/L) rapidly oxidizes hydrogen sulfide during emergency situations. Liming increases pH, shifting equilibrium away from toxic H_2S form and precipitating sulfides as insoluble metal sulfides. Reduction of organic inputs through controlled feeding and removal of dead fish and vegetation prevents organic accumulation.

Biological Parameters of Water Quality

Biological components of water quality include planktonic communities, bacterial populations, and presence of potentially harmful organisms.

Phytoplankton

Phytoplankton comprises microscopic algae suspended in water column, forming the foundation of aquatic food webs. Phytoplankton provides multiple benefits in pond aquaculture including oxygen production through photosynthesis, natural food for filter-feeding fish and zooplankton, turbidity creation that reduces rooted weed growth, and indication of pond fertility and productivity.

Major phytoplankton groups in freshwater ponds include green algae (Chlorophyceae), diatoms (Bacillariophyceae), blue-green algae or cyanobacteria, and dinoflagellates. Green algae and diatoms are generally desirable, producing oxygen efficiently and serving as quality food sources. Diatoms particularly indicate healthy, stable pond conditions.

Cyanobacteria (blue-green algae) present management challenges despite oxygen production. Many cyanobacterial species produce toxins including hepatotoxins affecting liver, neurotoxins causing paralysis, and dermatotoxins irritating skin and gills. Toxic blooms cause fish mortality and render water unsuitable for culture. Cyanobacteria dominate in ponds with high nitrogen:phosphorus ratios, elevated pH, calm conditions, and excessive nutrient loading.

Phytoplankton density influences multiple water quality parameters. Optimal densities (Secchi depth 30-40 cm, chlorophyll-a 30-80 µg/L) provide steady oxygen production without excessive instability. Dense blooms (Secchi depth <20 cm) create diurnal oxygen extremes and face collapse risks. Sudden phytoplankton die-offs or "bloom crashes" release ammonia and organic matter, consuming oxygen and potentially causing mass fish mortality.

Management of phytoplankton involves maintaining moderate fertility through controlled fertilization, avoiding excessive nitrogen inputs favoring cyanobacteria, promoting water circulation to prevent surface scum formation, and stocking appropriate densities of filter-feeding fish (silver carp, bighead carp) that graze on phytoplankton. Copper sulfate application at 0.1-0.5 mg/L selectively suppresses cyanobacteria, though it should be used judiciously due to potential copper toxicity and impacts on desirable algae.

Zooplankton

Zooplankton comprises microscopic animals feeding primarily on phytoplankton and bacteria, serving as critical food sources for fish larvae and juveniles. Major freshwater zooplankton groups include rotifers (50-200 µm), cladocerans or water fleas (0.2-3 mm), and copepods (0.5-2 mm).

Rotifers are highly abundant, reproduce rapidly, and represent the primary food for first-feeding fish larvae due to appropriate size and slow swimming. Rotifer densities of 50-200 per liter are optimal for larval rearing systems. **Cladocerans** including Daphnia, Moina, and Ceriodaphnia serve as excellent food for fingerlings, containing 50-60% protein and highly digestible nutrients. **Copepods** are nutritious but more difficult for fish to capture due to rapid escape responses, and some species prey upon fish larvae.

Zooplankton production in ponds depends on phytoplankton availability, with peak zooplankton densities typically occurring 1-2 weeks after phytoplankton blooms. Organic fertilization favors bacterial development supporting rotifer and zooplankton production. Maintaining diverse zooplankton communities provides stable natural food supply for stocked fish. Excessive filter-feeding fish densities can eliminate zooplankton through overgrazing, necessitating balanced stocking strategies.

Pathogenic Organisms

Water serves as the medium for disease transmission in aquaculture, harboring various pathogens including bacteria, viruses, fungi, and parasites. **Pathogenic bacteria** like Aeromonas, Pseudomonas, Edwardsiella, and Flavobacterium occur naturally in aquatic environments, becoming problematic when fish are stressed or immune-compromised by poor water quality.

Viral pathogens require living host cells for replication and can persist in water, sediments, and carrier fish. Major viral diseases including infectious hematopoietic necrosis (IHN), spring viremia of carp (SVC), and infectious pancreatic necrosis (IPN) cause significant losses in susceptible populations.

Parasitic organisms including protozoans (Ichthyophthirius, Trichodina, Costia), monogenean flukes, copepods (Lernaea, Argulus), and leeches commonly infest fish in pond culture systems. Parasite burdens increase with poor water quality, overcrowding, and nutritional stress.

Managing biological water quality requires integrated approaches including maintaining optimal physical and chemical parameters to support fish immunity, avoiding overcrowding that facilitates disease transmission, implementing biosecurity protocols preventing pathogen introduction, quarantining new fish stocks, disinfecting equipment, and periodic pond drying and liming to reduce pathogen populations. Early detection through regular fish health monitoring enables prompt treatment before disease outbreaks escalate.

Water Quality Monitoring and Testing

Systematic monitoring of water quality parameters enables early detection of problems and timely corrective actions. Monitoring frequency depends on culture intensity, species sensitivity, and season. **Intensive culture systems** require multiple daily measurements of dissolved oxygen and temperature, with weekly or biweekly assessment of other parameters. **Semi-intensive ponds** need daily morning dissolved oxygen checks, weekly pH and temperature readings, and monthly comprehensive testing. **Extensive systems** may adequately monitor with weekly dissolved oxygen and fortnightly comprehensive assessments.

Field testing equipment includes portable dissolved oxygen meters using optical or electrochemical sensors (accuracy ±0.1 mg/L), portable pH meters or colorimetric test kits (accuracy ±0.1 pH units), thermometers (accuracy ±0.5°C), Secchi disks for transparency, and test kits for ammonia, nitrite, alkalinity, and hardness using colorimetric methods. Regular calibration of meters using standard solutions ensures measurement accuracy.

Laboratory analysis provides more comprehensive assessment of parameters difficult to measure in field conditions, including specific nitrogen species, heavy metals, pesticides, biochemical oxygen demand, total suspended solids, and microbial counts. Accredited water testing laboratories should be utilized periodically to validate field measurements and conduct specialized analyses.

Record keeping of water quality data enables identification of trends, seasonal patterns, and relationships between parameters and fish performance. Data should be recorded systematically in logbooks or digital systems, noting date, time, location, weather conditions, and all measured parameters. Graphing data over time reveals patterns not apparent in individual measurements. Statistical analysis can identify correlations between water quality and production metrics like growth rates, feed conversion, and mortality.

Water Quality Management Strategies

Effective water quality management integrates preventive practices, routine monitoring, and corrective interventions.

Water Exchange and Flow-Through Systems

Water exchange dilutes accumulated metabolites, replenishes oxygen, and moderates temperature in culture ponds. Flow-through systems provide continuous water replacement, while static ponds employ batch or periodic exchange. The appropriate exchange rate depends on stocking density, feeding rate, and source water quality.

Low-intensity culture in fertilized ponds may require minimal exchange of 2-5% weekly, primarily compensating for evaporation and seepage. **Semi-intensive feeding systems** typically need 5-15% daily exchange to remove metabolites while retaining plankton productivity. **Intensive systems** without supplemental aeration may require 10-30% daily exchange, though this becomes economically and environmentally unsustainable at large scales.

Flow-through systems for salmonids and other species requiring cold, highly oxygenated water provide complete water replacement multiple times hourly. Water velocities of 10-30 cm/second maintain self-cleaning conditions and exercise fish. However, flow-through systems require abundant, high-quality water sources and face increasing environmental scrutiny due to effluent discharge impacts.

Water exchange timing influences effectiveness. **Afternoon exchange** during peak photosynthetic oxygen production minimizes disruption of oxygen dynamics. **Nighttime exchange** can provide oxygen supplementation if source water contains adequate DO. However, nighttime exchange with poor-quality source water may worsen conditions. Avoiding exchange during predawn hours when pond oxygen is naturally lowest prevents sudden water quality deterioration.

Source water quality must be assessed before exchange, as poor-quality source water can deteriorate pond conditions. Groundwater may require aeration to strip excess CO_2 and increase oxygen before entry. Surface water should be screened to exclude wild fish, predators, and disease vectors. During algal blooms or pollution events in source water bodies, exchange should be suspended until conditions improve.

Aeration and Circulation

Mechanical aeration represents the most effective method for increasing dissolved oxygen and supporting intensive aquaculture. Aerators function through two mechanisms: increasing air-water surface contact area to enhance atmospheric oxygen absorption, and creating water circulation that distributes oxygen throughout the pond.

Paddlewheel aerators consist of rotating blades mounted on a horizontal shaft, splashing water vigorously and creating horizontal circulation. Paddlewheel aerators exhibit high oxygen transfer efficiency of 1.5-2.5 kg O_2 per kilowatt-hour, making them the most popular choice for pond aquaculture. Typical placement is along pond edges or on floats, with spacing of 50-100 meters to ensure coverage of entire pond area.

Vertical pump aerators or propeller-aspirator-pump aerators propel water upward through a hollow shaft that aspirates air, mixing air and water and discharging the mixture as a fountain spray. These provide good oxygen transfer (1.0-1.8 kg O_2/kWh) and excellent circulation, drawing water from deeper layers to the surface for reoxygenation. They are particularly effective in deep ponds prone to stratification.

Diffused aeration systems introduce compressed air through diffusers placed at pond bottom, with rising air bubbles transferring oxygen and creating vertical circulation. While diffused aeration provides excellent circulation and destratification, oxygen transfer efficiency is relatively low (0.8-1.5 kg O_2/kWh) due to short bubble contact time. These systems require air compressors and extensive piping infrastructure, increasing capital and maintenance costs.

Aspirator-pump aerators spray water into the air without mechanical splashing, relying on droplet formation to increase surface area for oxygen absorption. These exhibit moderate oxygen transfer efficiency (1.2-1.8 kg O_2/kWh) and lower energy consumption than paddlewheels but provide less circulation.

Aerator operation strategies balance oxygen provision with energy costs. **Emergency aeration** activates when dissolved oxygen falls below critical levels (2-3 mg/L), operating continuously until oxygen recovers above 4-5 mg/L. **Preventive aeration** begins earlier (3-4 mg/L) and operates less intensely, reducing stress on fish while consuming less energy.

Scheduled aeration operates during predictable low-oxygen periods (typically midnight to dawn), anticipating oxygen depletion before it becomes critical.

The required aeration capacity depends on fish biomass, feeding rate, and temperature. A general guideline suggests:

Aerator HP = (Biomass in kg × Feeding rate in kg/day × 0.001) / AOT

Where AOT is aerator oxygen transfer rate in kg O_2/HP/hour (typically 1.5-2.0 for paddlewheels). For example, a pond with 5,000 kg fish biomass fed at 50 kg/day requires approximately 0.03 to 0.05 HP per kg biomass, or 150-250 HP total for emergency conditions.

Fertilization Management

Pond fertilization stimulates phytoplankton productivity, providing natural food and oxygen production. Fertilization strategies employ organic fertilizers, inorganic fertilizers, or combinations thereof.

Organic fertilizers including animal manures, composts, and agricultural byproducts supply nutrients gradually through decomposition while stimulating bacterial populations that support zooplankton production. Common organic fertilizers include cattle manure (100-300 kg/ha/week), chicken manure (50-150 kg/ha/week), and pig manure (80-200 kg/ha/week). Organic fertilization creates complex food webs benefiting omnivorous and filter-feeding species but increases oxygen demand and sediment accumulation.

Inorganic fertilizers provide immediately available nutrients, stimulating phytoplankton growth rapidly without contributing organic matter. Nitrogen and phosphorus represent the primary limiting nutrients in freshwater systems, with typical application rates of 20-40 kg N/ha and 5-15 kg P/ha. Common formulations include urea (46% N), ammonium sulfate (21% N), diammonium phosphate (18% N, 46% P_2O_5), and triple superphosphate (46% P_2O_5).

The ideal **nitrogen:phosphorus ratio** for freshwater phytoplankton production approximates 7-10:1 by weight. Deviation from optimal ratios causes nutrient imbalance, with excess nitrogen favoring cyanobacterial blooms and excess phosphorus limiting productivity. Regular monitoring of phytoplankton density and composition guides fertilization adjustments.

Fertilization should be suspended when Secchi depth falls below 20 cm, during hot weather with poor oxygen dynamics, after fish show stress signs, or when cyanobacterial blooms develop. Resumption of fertilization after suspension should be gradual with reduced application rates.

Liming Programs

Liming corrects acidic conditions, increases alkalinity and hardness, improves nutrient availability, enhances decomposition of organic matter, and disinfects pond bottoms. Lime materials include agricultural limestone ($CaCO_3$), hydrated lime ($Ca(OH)_2$), and quicklime (CaO), each with distinct properties and applications.

Agricultural limestone (calcium carbonate) is the safest and most commonly used liming material, though it reacts slowly. Application rates of 250-2,000 kg/ha depend on soil pH and buffering capacity. Limestone can be applied to water or broadcast on dried pond bottoms, with bottom application providing longer-lasting effects.

Hydrated lime (calcium hydroxide) reacts more rapidly than limestone and provides stronger pH elevation, making it suitable for emergency pH correction. Application rates typically range from 100-500 kg/ha. Hydrated lime also exhibits disinfectant properties, killing parasites, bacteria, and disease organisms on pond bottoms.

Quicklime (calcium oxide) is the most reactive and caustic liming material, used primarily for pond bottom disinfection between crops rather than routine liming. Quicklime application at 1,000-2,500 kg/ha to dried pond bottoms eliminates pathogens, parasites, predators, and unwanted fish. The exothermic reaction when quicklime contacts water generates heat and extreme alkalinity (pH >12) that destroys biological organisms.

Liming frequency depends on soil and water chemistry. Ponds with pH consistently below 6.5 or alkalinity below 20 mg/L require regular liming at 3-6 month intervals. Well-buffered ponds with alkalinity above 50 mg/L and pH above 7.0 need infrequent or no liming. Pre-stocking liming of new ponds or renovated ponds is standard practice regardless of measured pH.

Probiotics and Water Conditioners

Probiotic applications introduce beneficial bacterial consortia to aquaculture systems, competing with pathogens, enhancing organic matter decomposition, improving water quality, and potentially stimulating fish immune responses. Commercial probiotic products typically contain Bacillus species, lactic acid bacteria, photosynthetic bacteria, or combinations thereof.

Application rates vary by product formulation but typically range from 0.1-1.0 kg/ha weekly. Probiotics require dissolved oxygen above 3 mg/L and pH 6.5-8.5 for effective establishment. While scientific validation of probiotic efficacy in extensive systems remains limited, increasing adoption reflects perceived benefits in intensive culture.

Chemical water conditioners include various compounds marketed for water quality improvement. **Zeolites** remove ammonia through ion exchange, providing temporary ammonia control at application rates of 50-200 kg/ha. **Activated carbon** adsorbs organic compounds, colors, and some toxins but is expensive for large-scale use. **Alum and gypsum** clarify turbid water through flocculation. **Commercial water conditioners** often contain proprietary blends of minerals, enzymes, and bacteria, with variable efficacy.

Critical evaluation of water conditioners should consider scientific evidence, cost-effectiveness compared to proven management practices, and whether they address symptoms rather than underlying causes of water quality problems.

Applied Implications for Water Quality Management

Successful pisciculture fundamentally depends on maintaining water quality within acceptable ranges for cultured species. This requires understanding complex interactions among physical, chemical, and biological parameters, systematic monitoring to detect problems early, and timely implementation of corrective measures. The most effective water quality management integrates preventive strategies rather than reactive crisis intervention. Matching stocking densities and feeding rates to the oxygen-producing and waste-assimilating capacity of culture systems, maintaining adequate alkalinity and pH buffering, preventing excessive organic matter accumulation, and ensuring reliable aeration backup represent fundamental practices supporting sustainable intensification. Investment in water quality monitoring equipment, training personnel in proper measurement and interpretation of parameters, and maintaining comprehensive records enable data-driven management decisions that optimize production while safeguarding fish welfare and environmental quality.

3.3 Soil Preparation and Fertilization

The pond bottom soil represents more than simply the physical foundation of aquaculture systems; it serves as a biogeochemical reactor influencing water quality, nutrient cycling, natural food production, and overall pond productivity. **Soil management** encompasses preparation of pond bottoms before stocking, maintenance of soil conditions during culture periods, and renovation between production cycles. Understanding soil chemistry, biology, and management principles enables culturists to optimize pond productivity while preventing problems associated with deteriorated bottom conditions.

Soil Characteristics and Their Significance

Soil properties fundamentally determine pond suitability, water retention capacity, nutrient availability, and biological productivity. Evaluation of soil characteristics should precede pond construction and continue throughout operational life.

Soil Texture and Water-Holding Capacity

Soil texture refers to the relative proportions of sand, silt, and clay particles, profoundly influencing water retention, permeability, and fertility. **Clay particles** (<0.002 mm diameter) are critical for water retention due to their small size, large surface area, and cohesive properties. **Silt particles** (0.002-0.05 mm diameter) provide moderate water retention and nutrient-holding capacity. **Sand particles** (0.05-2.0 mm diameter) offer minimal water retention and nutrient storage but provide drainage and structural stability.

Ideal pond soils contain **20-40% clay**, creating adequate impermeability with acceptable seepage rates below 10-15 mm/day while maintaining sufficient porosity for oxygen penetration and root growth of benthic plants. Clay-loam and silty-clay-loam soils represent optimal textures, balancing water retention with workability and fertility.

Sandy soils with clay content below 15% exhibit excessive seepage, requiring expensive pond lining or soil amendments. Heavily clayey soils (>60% clay) create drainage problems, develop deep cracks when dried, and may exhibit poor fertility due to nutrient fixation. Organic soils or muck soils containing >20% organic matter compress over time, causing pond subsidence and uneven bottoms.

Soil texture analysis through mechanical sieving and sedimentation determines particle size distribution. The **textural triangle** classification system assigns soil to textural classes based on percentages of sand, silt, and clay, facilitating objective soil evaluation.

Soil pH and Acidity

Soil pH governs nutrient availability, microbial activity, and chemical reactions occurring in pond sediments. Most aquaculture species thrive in ponds with soil pH ranging from 6.5 to 8.0. **Acid soils** with pH below 6.0 reduce bacterial decomposition of organic matter, decrease nutrient availability through aluminum and iron fixation, and may release toxic aluminum concentrations into overlying water.

Soil acidity originates from various sources including organic acid accumulation, aluminum and iron hydrolysis, and most problematically, oxidation of **acid sulfate materials**. Acid sulfate soils contain pyrite (FeS_2) formed under anaerobic, sulfate-rich conditions in marine or estuarine sediments. When exposed to oxygen during pond excavation or draining, pyrite oxidizes producing sulfuric acid:

$$4FeS_2 + 15O_2 + 14H_2O \rightarrow 4Fe(OH)_3 + 8H_2SO_4$$

This reaction can decrease soil pH to 2.5-4.0, generating extreme acidity that renders ponds unsuitable for aquaculture unless extensively ameliorated. Acid sulfate soil indicators include a sulfurous smell during excavation, yellow or red mottling in subsoils from iron oxidation, and persistent low pH despite liming.

Alkaline soils with pH above 8.5 may contain excessive sodium (sodic soils) or carbonates, creating unfavorable conditions. Sodic soils exhibit poor structure, impermeability to water and air, and elevated pH that can exceed 9.5. Highly calcareous soils with excess calcium carbonate may fix phosphorus as insoluble calcium phosphate, limiting availability for phytoplankton production.

Laboratory analysis of soil pH should be conducted using soil:water ratios of 1:2.5 or 1:5, with measurements taken on fresh samples and after oxidation. **Oxidized pH** is determined by air-drying soil samples for several weeks before measurement, revealing potential acidity that would develop in drained pond bottoms or newly excavated areas.

Soil Organic Matter

Organic matter in pond soils originates from uneaten feed, fish feces, dead plankton, macrophyte debris, and microbial biomass. Moderate organic matter content (2-4%) benefits pond productivity by slowly releasing nutrients, improving soil structure, increasing cation exchange capacity, and supporting diverse microbial communities.

Excessive organic matter accumulation (>8-10%) creates problems including oxygen depletion during decomposition, release of ammonia and hydrogen sulfide, reduced pond volume through sediment buildup, and instability of pond bottom conditions. Organic-rich sediments create an oxygen demand that can reach 2-5 g O_2/m^2/day, severely stressing fish during periods of low oxygen production.

The **carbon:nitrogen ratio** (C:N) of accumulated organic matter influences decomposition rates and nutrient release. Fresh organic matter with C:N ratios above 20:1 decomposes slowly and may immobilize nitrogen as microorganisms scavenge available nitrogen for their growth. Well-decomposed organic matter with C:N ratios below 12:1 releases nitrogen more readily, supporting plankton productivity.

Management of organic matter accumulation requires balanced feeding practices that minimize waste, periodic removal of sediments through dredging or excavation, and thorough drying and oxidation of pond bottoms between culture cycles.

Cation Exchange Capacity and Nutrient Retention

Cation exchange capacity (CEC) measures the soil's ability to hold and exchange positively charged ions (cations) including calcium, magnesium, potassium, ammonium, and sodium. CEC is expressed in milliequivalents per 100 grams of soil (meq/100g) or centimoles of charge per kilogram (cmol×/kg).

Clay minerals and organic matter provide the primary sources of CEC, with values typically ranging from 5-15 meq/100g for sandy soils, 15-30 meq/100g for loamy soils, and 30-60 meq/100g for clayey soils.

High CEC soils retain nutrients more effectively, buffering against nutrient loss through leaching and providing gradual nutrient release to support productivity.

The proportion of CEC occupied by different cations influences soil fertility and water chemistry. **Base saturation** refers to the percentage of CEC occupied by basic cations (calcium, magnesium, potassium, sodium) rather than acidic cations (hydrogen, aluminum). Optimal base saturation exceeds 80%, with calcium dominating (60-75% of CEC), magnesium at 10-20%, potassium at 2-5%, and sodium below 5%.

Soils with aluminum saturation exceeding 20% exhibit aluminum toxicity problems, while sodium saturation above 15% indicates sodic conditions with associated structural and pH problems.

Pond Bottom Preparation

Proper preparation of pond bottoms before stocking establishes favorable conditions for fish growth and natural food production while preventing problems during the culture period.

Draining and Drying

Complete draining of ponds between culture cycles represents the single most important pond management practice for semi-intensive and intensive systems. Draining removes accumulated metabolites, eliminates undesirable fish, controls predators and competitors, allows soil oxidation, and facilitates maintenance operations.

After harvest and complete drainage, the pond bottom should be exposed to **sun drying** for 2-4 weeks or until soil develops fine cracks 1-2 cm wide and 5-10 cm deep. Drying accomplishes multiple benefits including oxidation of reduced compounds (hydrogen sulfide, methane, reduced iron and manganese), mineralization of organic matter with nutrient release, elimination of anaerobic bacteria, killing of parasites and disease organisms, consolidation of soft sediments, and enhancement of soil structure.

The drying period required depends on climate, soil type, sediment thickness, and time of year. In tropical regions during hot, dry seasons, adequate drying may occur in 10-15 days. Cool, humid periods may require 4-6 weeks for comparable drying. Clayey soils require longer drying than sandy soils due to slower water loss.

Overdrying should be avoided, as excessively deep cracks (>5 cm wide) can persist after refilling, creating refuge areas for predatory insects and making harvesting difficult. In practice, drying to a depth where soil can be tilled or raked but maintains some moisture is optimal.

Mechanical tilling or plowing of dried pond bottoms to depths of 10-15 cm accelerates oxidation by increasing soil-air contact, breaks up compacted layers, incorporates organic matter, and facilitates lime and fertilizer mixing. Tilling should be followed by leveling to maintain design slopes and prevent formation of low areas that would hinder complete drainage in subsequent cycles.

Removal of Accumulated Sediments

Progressive accumulation of organic-rich sediments in pond bottoms reduces water volume, deteriorates bottom conditions, and increases oxygen demand. When sediment layers exceed 15-20 cm thickness, **sediment removal** becomes necessary.

Manual excavation using shovels and baskets, mechanical dredging using backhoes or draglines, or hydraulic dredging using suction pumps can remove accumulated sediments. Removed sediments should not be piled on pond embankments where they can cause structural problems, but should be disposed of in designated areas or applied to agricultural fields as organic amendments.

The frequency of sediment removal depends on culture intensity and organic loading. Extensively managed ponds may require sediment removal every 5-10 years, while intensive feeding operations may need removal every 2-3 years. Some highly intensive systems excavate 5-10 cm of bottom sediments annually during pond renovation.

Liming of Pond Bottoms

Application of lime to dried pond bottoms provides more effective and longer-lasting benefits than liming water surfaces. **Bottom liming** neutralizes soil acidity, increases pH and base saturation, enhances nutrient availability, improves soil structure, stimulates microbial decomposition, and disinfects pond bottoms.

The lime requirement depends on soil pH, buffer capacity, and organic matter content, determined through laboratory analysis. A simplified approach estimates lime requirement based on soil pH:

- Soil pH 4.0-5.0: 4,000-8,000 kg/ha $CaCO_3$ equivalent
- Soil pH 5.0-6.0: 2,000-4,000 kg/ha $CaCO_3$ equivalent
- Soil pH 6.0-6.5: 500-2,000 kg/ha $CaCO_3$ equivalent
- Soil pH 6.5-7.0: 250-500 kg/ha $CaCO_3$ equivalent

Agricultural limestone (calcium carbonate, $CaCO_3$) represents the standard liming material with 90-100% calcium carbonate equivalent (CCE), depending on purity. Application is made to dried pond bottoms and incorporated through tilling or raking for uniform distribution and faster reaction.

Dolomitic limestone ($CaMg(CO_3)_2$) provides both calcium and magnesium, beneficial in soils deficient in magnesium. The magnesium component reacts more slowly than calcium carbonate, providing prolonged buffering.

Hydrated lime ($Ca(OH)_2$) or slaked lime contains approximately 135% CCE and reacts more rapidly than agricultural limestone. Hydrated lime is particularly useful for emergency pH correction or when rapid alkalinity increase is needed. The application rate is calculated by:

Hydrated lime required = Limestone equivalent × (100/135)

Quicklime (CaO) contains approximately 180% CCE and reacts extremely rapidly and exothermically with water, generating intense heat and high pH. Quicklime is applied primarily for disinfection rather than routine liming. Application at 1,000-2,500 kg/ha to dried pond bottoms kills fish parasites, predatory insects, disease organisms, weed seeds, and unwanted fish eggs. The disinfection process requires the pond bottom to be dried until slightly moist (not saturated), lime broadcast uniformly, and then left for 7-10 days before filling.

Safety precautions are essential when handling quicklime, as it causes severe burns upon contact with skin or eyes and generates dangerous heat when water is added. Workers must wear protective equipment including gloves, boots, long-sleeved clothing, and eye protection.

Elimination of Predators and Competitors

Wild fish, amphibians, aquatic insects, and other organisms entering ponds compete with cultured species for food and oxygen, prey upon fry and fingerlings, or serve as disease vectors. Systematic elimination before stocking prevents these problems.

Complete pond draining removes most fish and aquatic organisms, though some species survive in moist bottom sediments or residual pools. **Sun drying** for 2-3 weeks kills most remaining organisms. **Quicklime application** provides additional disinfection, eliminating hardy species, insect larvae, and parasite stages.

Screening of inlet structures with fine mesh (2-5 mm) prevents entry of wild fish, insects, and predators during pond filling. Regular inspection and cleaning of screens maintains effectiveness. For ponds receiving water from rivers or lakes with abundant wild fish populations, more elaborate screening or filtration systems may be necessary.

Chemical eradication using piscicides eliminates fish populations in situations where draining is impossible. **Rotenone**, a natural plant-derived compound, is the most widely used piscicide at concentrations of 1-2 mg/L active ingredient. Rotenone selectively affects fish and aquatic insects by inhibiting cellular respiration while exhibiting low toxicity to mammals, birds, and most invertebrates. The compound degrades naturally within 2-4 weeks, after which ponds can be safely restocked.

Alternative piscicides include **tea seed cake** (saponin-containing compound) at 15-25 mg/L, effective against fish and predatory insects while supporting subsequent zooplankton development. Synthetic piscicides like antimycin are highly effective but face regulatory restrictions in many regions.

Pond Fertilization Principles and Practices

Pond fertilization stimulates primary productivity, establishing phytoplankton-based food webs that provide natural nutrition for cultured fish, produce oxygen through photosynthesis, create turbidity preventing rooted weed growth, and enhance overall pond productivity. Fertilization represents a fundamental management practice in semi-intensive aquaculture systems where natural food contributes significantly to fish nutrition.

Nutrient Limitation in Aquatic Systems

Aquatic primary production is typically limited by availability of nitrogen and phosphorus, which are required in relatively large quantities but often occur at low concentrations in natural waters. **Liebig's Law of the Minimum** states that productivity is constrained by the nutrient present in the most limiting supply relative to the demands of organisms.

In most freshwater systems, **phosphorus** represents the primary limiting nutrient for phytoplankton production. Phosphorus inputs from atmospheric deposition and erosion are minimal compared to biological demands. Phosphorus also binds readily to soil particles, precipitates with calcium in alkaline waters, and accumulates in sediments, limiting availability in the water column.

Nitrogen limitation can occur in ponds with high phosphorus availability, particularly when nitrogen inputs from feed and fish excretion are insufficient to match phytoplankton demands. However, nitrogen-fixing cyanobacteria can exploit atmospheric nitrogen, partially alleviating pure nitrogen limitation though potentially creating water quality problems.

The **optimal N:P ratio** for freshwater phytoplankton approximates 7-10:1 by weight (16:1 by atoms), based on average phytoplankton composition. Fertilization programs should maintain this ratio to prevent nutrient imbalance. Excess nitrogen relative to phosphorus (N:P >15:1) favors cyanobacterial blooms, while excess phosphorus (N:P <5:1) limits productivity despite adequate phosphorus.

Types of Fertilizers

Pond fertilizers are classified as organic fertilizers derived from plant or animal sources and inorganic fertilizers comprising pure mineral compounds.

Organic Fertilizers

Organic fertilizers provide nutrients gradually through microbial decomposition while supplying carbon compounds supporting bacterial growth, which forms the base of detrital food webs. Organic fertilization creates complex, stable food webs with multiple trophic levels including bacteria, protozoa, rotifers, and crustacean zooplankton, benefiting omnivorous fish species that can exploit diverse food sources.

Common organic fertilizers include animal manures, composts, agricultural residues, and industrial organic byproducts. **Poultry manure** contains relatively high nitrogen (3-4% N, 2-3% P_2O_5, 1-2% K_2O on dry weight basis) and decomposes rapidly, making it an excellent pond fertilizer at application rates of 50-150 kg dry weight per hectare per week. Fresh poultry manure can be applied directly, though composting for 2-4 weeks reduces oxygen demand and eliminates pathogens.

Cattle manure contains lower nutrient concentrations (0.5-1.5% N, 0.3-1.0% P_2O_5) and decomposes more slowly due to higher fiber content. Application rates typically range from 100-300 kg/ha/week. Cattle manure supports gradual nutrient release and stimulates bacterial production effectively.

Pig manure exhibits intermediate composition (2-3% N, 1-2% P_2O_5) and decomposition rates, applied at 80-200 kg/ha/week. Pig manure is particularly effective for stimulating zooplankton production.

Fresh animal manures should be used cautiously as they contain readily decomposable organic matter that consumes oxygen. **Composting** for 3-4 weeks stabilizes organic matter, reduces oxygen demand, eliminates parasites and disease organisms, and improves handling characteristics. Composted manures can be applied at higher rates with less risk of oxygen depletion.

Green manures from terrestrial or aquatic plants provide carbon and nutrients while avoiding disease transmission risks associated with animal wastes. Common green manure materials include alfalfa, rice straw, water hyacinth, and aquatic weeds. Application rates of 500-2,000 kg/ha fresh weight supply nutrients and support microbial food webs, though decomposition is slower than animal manures due to higher C:N ratios.

Sewage sludge or biosolids from municipal wastewater treatment provide nutrient-rich organic fertilizer, though use raises concerns about pathogens, heavy metals, and emerging contaminants. Only properly treated and tested biosolids should be considered, and use for food fish production faces regulatory and consumer acceptance challenges.

Organic fertilization offers several advantages including slow-release nutrient supply, stimulation of bacterial and detrital food webs, improvement of soil structure and fertility, and low cost where materials are locally available. However, organic fertilization also presents challenges including high labor requirements for collection and application, variable and often low nutrient concentrations, introduction of excessive organic matter causing oxygen depletion, potential disease transmission, and difficulty in precise nutrient management.

Inorganic Fertilizers

Inorganic fertilizers supply specific nutrients in immediately available forms, enabling precise control of nutrient inputs and rapid stimulation of phytoplankton growth. Inorganic fertilization typically targets nitrogen and phosphorus, the primary limiting nutrients.

Nitrogen fertilizers commonly used in aquaculture include:

- **Urea** ($CO(NH_2)_2$, 46% N): Most concentrated nitrogen source, highly soluble, rapidly available. Urea hydrolyzes to ammonia, which may elevate ammonia concentrations transiently but is quickly assimilated by phytoplankton. Application rate: 20-40 kg N/ha/application.
- **Ammonium sulfate** ($(NH_4)_2SO_4$, 21% N, 24% S): Provides both nitrogen and sulfur, acidifying effect may benefit alkaline soils. Lower nitrogen concentration requires larger application quantities. Application rate: 20-40 kg N/ha/application.
- **Ammonium nitrate** (NH_4NO_3, 34% N): Supplies nitrogen in both ammonium and nitrate forms for rapid availability. Use may be restricted due to explosive properties.

Phosphorus fertilizers include:

- **Triple superphosphate** (TSP, 45-46% P_2O_5): Concentrated phosphorus source, moderately water-soluble. Application rate: 5-15 kg P_2O_5/ha/application.
- **Diammonium phosphate** (DAP, 18% N, 46% P_2O_5): Supplies both nitrogen and phosphorus in balanced ratio for freshwater systems. Highly soluble and rapidly available. Application rate: 5-15 kg P_2O_5/ha (simultaneously providing 2-6 kg N/ha).
- **Single superphosphate** (SSP, 16-18% P_2O_5): Lower concentration phosphorus source also providing calcium and sulfur. Application rate: 5-15 kg P_2O_5/ha.

Application rates of inorganic fertilizers depend on pond productivity, stocking density, feeding practices, and season. Initial fertilization at pond filling typically employs higher rates (30-50 kg N/ha and 10-20 kg P/ha) to establish plankton blooms rapidly. Maintenance fertilization at weekly or biweekly intervals uses lower rates (15-30 kg N/ha and 5-10 kg P/ha) to sustain moderate productivity.

Fertilizer application frequency is guided by water transparency monitoring. When Secchi depth exceeds 40-50 cm, indicating declining plankton density, fertilization should be applied. When Secchi depth falls below 25 cm, fertilization should be withheld until water clears slightly to prevent excessive blooms.

Combined fertilization using both organic and inorganic fertilizers exploits advantages of each approach. A common strategy applies organic fertilizers at moderate rates to establish bacterial food webs while supplementing with inorganic fertilizers to maintain phytoplankton productivity. This combination creates diverse, stable food webs producing both zooplankton and phytoplankton for fish consumption.

Application Methods

Broadcasting represents the simplest fertilization method, with solid fertilizers spread uniformly across the pond surface from a boat or thrown from pond banks. Fertilizers should be broadcast during calm periods to achieve uniform distribution and should be applied to multiple locations for large ponds.

Platform placement involves constructing submerged platforms or suspended bags containing fertilizers, allowing gradual dissolution and localized nutrient release. This method reduces nutrient loss to sediments and extends fertilizer effectiveness, though it requires more labor and materials for platform construction.

Liquid fertilization dissolves soluble fertilizers in water and sprays the solution across pond surfaces, ensuring uniform distribution. Liquid application is particularly suitable for large ponds where broadcast distribution is difficult.

Pre-mixing fertilizers into pond water before filling or preparing fertilizer solutions in advance may improve dissolution and availability, though additional labor is required.

Fertilization Management Strategies

Effective fertilization programs integrate biological understanding with practical monitoring to optimize productivity while avoiding problems associated with excessive fertility.

Initial Fertilization

Pre-stocking fertilization establishes plankton blooms before fingerling introduction, ensuring adequate natural food availability immediately upon stocking. Ponds should be filled and fertilized 7-14 days before stocking to allow phytoplankton development. Initial fertilization employs higher nutrient rates to overcome sorption to fresh soil particles and stimulate bloom formation.

A typical pre-stocking protocol applies 300-500 kg/ha composted organic manure combined with 30-40 kg N/ha and 10-15 kg P_2O_5/ha from inorganic sources. This heavy initial dose jumpstarts microbial and planktonic communities. Water transparency should reach 25-35 cm Secchi depth before fingerlings are introduced, indicating moderate plankton density appropriate for supporting larval and juvenile fish.

Maintenance Fertilization

Following stocking, **maintenance fertilization** sustains plankton productivity at moderate levels throughout the culture period. Application frequency typically follows weekly or biweekly schedules, with rates adjusted based on water transparency monitoring.

A standard maintenance program applies:

- Organic fertilizer: 100-200 kg/ha/week
- Nitrogen: 15-25 kg N/ha/week
- Phosphorus: 5-10 kg P_2O_5/ha/week

These rates maintain Secchi depths of 25-40 cm, providing adequate natural food and oxygen production without risking bloom crashes. Application should occur during morning hours when dissolved oxygen is rising, avoiding late afternoon or evening when baseline oxygen may already be declining.

Adjustment Based on Feeding Intensity

As culture progresses and fish size increases, supplemental feeding typically intensifies, reducing reliance on natural food and increasing nutrient inputs from uneaten feed and fish excretion. Fertilization rates should be reduced proportionally to prevent over-enrichment.

When feeding rates exceed 30-40 kg feed/ha/day, maintenance fertilization can often be discontinued entirely, as nutrients from feeding operations alone sustain moderate plankton productivity. Continuing heavy fertilization under intensive feeding regimens creates excessive plankton blooms, exacerbating oxygen fluctuations and instability.

Seasonal Adjustments

Phytoplankton productivity varies seasonally with light intensity, day length, and temperature. In tropical regions, productivity peaks during hot seasons with high solar radiation, requiring reduced fertilization to prevent excessive blooms. During cooler seasons or monsoon periods with reduced sunlight, higher fertilization rates may be necessary to maintain adequate productivity.

Temperate regions exhibit pronounced seasonal patterns, with vigorous spring and summer productivity but minimal winter production. Fertilization programs should reflect these patterns, intensifying during productive seasons and suspending during winter when water temperatures fall below 15°C and phytoplankton growth becomes negligible.

Suspension of Fertilization

Fertilization should be temporarily suspended under several circumstances:

- Water transparency (Secchi depth) falls below 20 cm, indicating excessive plankton density
- Dissolved oxygen falls below 4 mg/L during morning measurements
- Fish exhibit stress behaviors suggesting oxygen deficiency
- Cyanobacterial blooms develop, identified by blue-green surface scums
- Heavy rainfall causes sudden water quality changes

- Water temperature exceeds 32-33°C, increasing oxygen demand
- Disease outbreaks occur, as fertilization may stress fish further

After conditions stabilize, fertilization can resume gradually at reduced rates, monitoring fish behavior and water quality closely.

Soil-Water Interactions and Nutrient Cycling

The pond bottom soil and overlying water column interact continuously through nutrient exchange, affecting productivity and water quality.

Nutrient Release from Sediments

Pond bottom sediments accumulate organic matter from uneaten feed, feces, dead plankton, and other sources, creating a nutrient reservoir that exchanges with the water column. **Mineralization** of organic matter by heterotrophic bacteria releases inorganic nutrients including ammonia-nitrogen, phosphate, carbon dioxide, and various minerals.

The rate of nutrient release from sediments depends on temperature, oxygen availability, organic matter content, microbial activity, and sediment composition. Under aerobic conditions, decomposition proceeds efficiently with minimal toxic byproduct formation. The relationship between temperature and mineralization rate follows:

$$R(T) = R(20) \times Q_{10}\wedge((T-20)/10)$$

Where R(T) is the mineralization rate at temperature T, R(20) is the rate at 20°C, and Q_{10} typically ranges from 2.0 to 3.0 for organic matter decomposition.

Sediment nutrient release can contribute 30-60% of total nitrogen and phosphorus supporting phytoplankton production in fertilized ponds. This internal loading continues even when external fertilization is suspended, explaining why phytoplankton persists in ponds with accumulated sediments despite ceased fertilization.

Sediment Oxygen Demand

Sediment oxygen demand (SOD) represents the oxygen consumption rate by bottom sediments, resulting from aerobic decomposition of organic matter and chemical oxidation of reduced compounds. SOD typically ranges from 0.5 to 5.0 g O_2/m^2/day, depending on organic loading, temperature, and sediment characteristics.

The SOD can be a major component of total pond oxygen consumption, particularly in intensively fed ponds with heavy sediment accumulation. For example, a pond with 2.0 g O_2/m^2/day SOD consumes 20 kg oxygen per hectare per day solely from bottom processes, equivalent to the respiratory demand of 400-500 kg of fish.

Managing sediment oxygen demand requires controlling organic inputs through efficient feeding, periodic sediment removal, and maintaining oxidizing conditions at the sediment-water interface through circulation and aeration. Complete pond draining and drying oxidizes accumulated organic matter, substantially reducing SOD in subsequent culture periods.

Phosphorus Dynamics and Sediment Binding

Phosphorus exhibits strong affinity for soil particles, binding through multiple mechanisms including adsorption onto clay minerals and iron/aluminum oxides, precipitation as calcium phosphates in alkaline conditions, and incorporation into organic matter. These binding processes reduce phosphorus availability in the water column, explaining why phosphorus often limits productivity despite substantial fertilization.

The **phosphorus binding capacity** depends on soil mineralogy, pH, iron and aluminum content, and calcium concentration. Soils rich in iron and aluminum oxides or amorphous clays exhibit high phosphorus retention. In acidic soils (pH <6.0), phosphorus binds to iron and aluminum compounds, while in alkaline soils (pH >7.5), phosphorus precipitates as calcium phosphates. Maximum phosphorus availability typically occurs at pH 6.5-7.5.

Phosphorus bound to sediments can be released back to the water column under specific conditions. **Redox potential** governs phosphorus release, with anoxic sediments releasing substantially more phosphorus than oxidized sediments. Under anaerobic conditions, iron (III) is reduced to iron (II), releasing associated phosphates. This internal phosphorus loading contributes to sudden algal blooms following periods of bottom water anoxia or sediment disturbance.

Liming increases phosphorus availability in acid soils by raising pH toward the optimal range and reducing iron/aluminum binding. However, excessive liming in already alkaline soils can decrease phosphorus availability through calcium phosphate precipitation. Organic matter additions improve phosphorus availability by forming soluble organo-phosphorus complexes and coating mineral surfaces, reducing phosphorus adsorption sites.

Nitrogen Cycling in Pond Systems

Nitrogen cycles through multiple chemical forms and biological transformations in aquaculture ponds. Fish excrete nitrogen primarily as ammonia, contributing 20-30 g ammonia-nitrogen per kilogram feed applied in intensive systems. This ammonia enters various nitrogen cycle pathways.

Ammonia assimilation by phytoplankton represents a major nitrogen removal pathway, with algae incorporating ammonia into proteins and other organic compounds. Ammonia uptake rates by phytoplankton can reach 50-100 kg N/ha/day during productive periods, providing significant water quality benefit while producing fish food.

Nitrification converts ammonia sequentially to nitrite and nitrate under aerobic conditions as previously described. Nitrification requires oxygen (4.6 kg O_2 per kg NH_3-N oxidized to NO_3-N), contributes to acidification through proton release, and depends on adequate alkalinity for optimal bacterial function.

Denitrification converts nitrate to nitrogen gas under anaerobic conditions in bottom sediments:

$2NO_3$✕ + Organic matter → $N_2 + CO_2 + H_2O$

Denitrification represents the primary nitrogen removal pathway from aquaculture systems, permanently eliminating nitrogen as gaseous N_2. Denitrification rates in pond sediments typically range from 10 to 50 kg N/ha/day, depending on nitrate availability, organic matter content, and extent of anoxic sediments.

Ammonia volatilization occurs when pH exceeds 8.5, with un-ionized ammonia escaping across the water surface. Volatilization losses typically account for 5-15% of total ammonia-nitrogen in alkaline ponds with vigorous aeration or wind mixing.

Understanding nitrogen cycling enables optimization of pond management to minimize toxic nitrogen accumulation while maintaining adequate nitrogen for phytoplankton production.

Soil Testing and Analysis

Regular soil testing provides objective information guiding fertilization decisions, identifying developing problems, and evaluating management effectiveness.

Sampling Procedures

Representative soil samples should be collected from multiple locations (8-12 points) across the pond bottom, avoiding areas near inlet/outlet structures or unusual features. Sampling depth should be 0-10 cm to represent the biologically active surface layer. Samples from multiple locations should be thoroughly mixed to create a composite sample of 0.5-1.0 kg for laboratory analysis.

Soil sampling should occur during pond draining after harvest, when the bottom is accessible but not completely dry. Sampling dried pond bottoms provides information relevant to lime requirement determination, while sampling moist bottoms better represents conditions during the culture period.

Laboratory Analysis Parameters

Comprehensive soil analysis should include:

- **Mechanical composition** (sand, silt, clay percentages) - indicates water-holding capacity and fertility potential
- **pH** - measured in water suspension and after oxidation (peroxide treatment) to reveal potential acidity
- **Organic matter content** - guides management of sediment accumulation and organic loading
- **Cation exchange capacity** - indicates nutrient retention capacity
- **Exchangeable cations** (Ca, Mg, K, Na, Al) - reveals nutrient status and potential toxicity problems
- **Available phosphorus** - extracted using appropriate methods (Olsen for alkaline soils, Bray for acid soils) indicates phosphorus fertility
- **Total nitrogen** - indicates nitrogen reservoir in organic matter
- **Extractable iron and aluminum** - high levels suggest phosphorus binding and potential toxicity in acid soils
- **Lime requirement** - determined by buffer pH or direct titration methods

Laboratory analysis costs typically range from $30-80 per sample for comprehensive testing, representing a minor investment relative to the value of improved management decisions.

Interpretation and Management Adjustments

Soil test results should be interpreted considering species requirements, management intensity, and economic constraints. General guidelines suggest:

- Soil pH 6.5-7.5: Optimal, no lime needed unless alkalinity is low
- Soil pH 6.0-6.5: Beneficial to apply 500-1,000 kg/ha limestone
- Soil pH <6.0: Apply 1,000-2,000+ kg/ha limestone based on buffer pH
- Organic matter >8%: Consider sediment removal or reduced feeding
- Available phosphorus <10 mg/kg: Increase phosphorus fertilization
- Available phosphorus >50 mg/kg: Reduce phosphorus inputs
- Exchangeable aluminum >2.0 meq/100g: Lime heavily to reduce aluminum solubility

Soil testing should be repeated annually or biannually to track long-term trends and verify that management practices maintain soil quality within acceptable ranges.

Problematic Soil Conditions and Remediation

Certain soil conditions create serious management challenges requiring specific remediation approaches.

Acid Sulfate Soils

Acid sulfate soils containing pyrite minerals represent one of the most challenging soil problems in aquaculture. Oxidation of pyrite during pond construction or draining generates sulfuric acid, potentially reducing pH to 2.5-4.0 and releasing toxic aluminum and iron concentrations.

Identification of potential acid sulfate soils before development is critical. Indicators include coastal or estuarine location, yellow or red subsoil mottling, sulfurous odor during excavation, persistent low pH despite liming, and characteristic vegetation (e.g., Melaleuca species). Laboratory testing for peroxide-oxidizable sulfur content provides quantitative assessment, with levels exceeding 0.03% sulfur (equivalent to 0.1% pyrite) indicating problematic conditions.

Remediation of acid sulfate soils requires extensive liming to neutralize generated acidity. Lime requirements often exceed 5-20 tons/ha of agricultural limestone or 3-15 tons/ha of quicklime. The lime requirement can be estimated from:

$$\text{Lime (kg/ha)} = \text{\% Pyrite} \times 31 \times 10{,}000$$

Where 31 represents the $CaCO_3$ equivalent needed to neutralize acid from 1 kg pyrite.

Management strategies for acid sulfate soils include:

Pre-flooding liming: Apply heavy lime doses to dry pond bottoms, allow oxidation and reaction for 2-3 months, flush repeatedly with freshwater to remove acid and soluble iron/aluminum, repeat liming, and then finally fill for culture.

Minimal disturbance: Construct ponds with minimal excavation to avoid exposing deep pyritic layers. Maintain water cover year-round to prevent oxidation, avoiding complete draining that would re-oxidize sediments.

Regular monitoring: Check pH and dissolved metals weekly, applying supplemental lime to maintain pH above 6.5. Water exchange should be minimized to prevent flushing added alkalinity.

Despite intensive management, acid sulfate soils generally remain problematic for aquaculture, and site selection should avoid such areas when possible.

Sodic Soils

Sodic soils containing excessive exchangeable sodium (>15% of CEC) exhibit poor structure, impermeability, and elevated pH (often >9.0). Sodium causes clay particle dispersion, resulting in surface crusting when dry and sticky, impermeable conditions when wet.

Sodic soil indicators include black surface crusts, columnar soil structure with cracks, poor water infiltration, and elevated pH with relatively low total alkalinity. Laboratory testing reveals high exchangeable sodium percentage (ESP >15%) and sodium adsorption ratio (SAR >13).

Remediation requires replacing exchangeable sodium with calcium through application of **gypsum** (calcium sulfate):

Gypsum required (kg/ha) = ESP × CEC × 1,720

Where ESP is expressed as a percentage, CEC in meq/100g, and 1,720 is a conversion factor.

Gypsum application at 2-5 tons/ha to dried pond bottoms, followed by leaching with fresh water to remove displaced sodium, gradually improves soil structure. Repeated applications over multiple years may be necessary for severely sodic conditions. Organic matter additions through manuring or green manure incorporation also improve sodic soil structure.

Highly Permeable Soils

Sandy soils with excessive permeability (seepage >25 mm/day) cannot retain water economically without treatment. Several approaches reduce seepage:

Bentonite application: Sodium bentonite clay swells upon wetting, filling soil pores and reducing permeability. Application rates of 1-3 kg/ m^2 (10,000-30,000 kg/ha), spread uniformly and compacted into the top 10-15 cm of soil, can reduce seepage by 70-90%. However, bentonite is expensive ($200-500/ton) and treatment costs may exceed $20,000-60,000/ha.

Soil mixing: Incorporating clayey soil into sandy pond bottoms by spreading a 15-20 cm clay layer and mixing it thoroughly with underlying sand can improve water retention if suitable clay sources are available nearby.

Synthetic liners: Polyethylene, PVC, or EPDM membrane liners provide complete impermeability but entail high costs ($5-20/$m^2$ installed) and restrict bottom-dwelling organisms. Liners are economical primarily for small, high-value operations.

Dispersed clay treatment: Application of sodium-based dispersants like soda ash (sodium carbonate) at 100-200 kg/ha causes clay particle dispersion, filling pores and reducing seepage. This approach is less reliable than bentonite and may create subsequent sodicity problems.

Applied Implications for Soil Management

Proper soil preparation and management represent foundational practices supporting productive, sustainable aquaculture. Systematic evaluation of soil characteristics before pond construction prevents future problems more economically than post-construction remediation. Regular pond draining and drying between culture cycles oxidizes accumulated organic matter, mineralizes nutrients, eliminates parasites, and regenerates soil conditions supporting subsequent production cycles. Judicious fertilization matched to pond productivity potential and adjusted for supplemental feeding inputs maintains moderate plankton productivity supporting natural food production without creating instability or excessive sediment accumulation. Long-term sustainability requires balancing nutrient inputs with removal pathways, periodically removing accumulated sediments when necessary, and maintaining soil pH and structure through appropriate liming programs. Investment in soil testing provides objective data enabling informed management decisions that optimize productivity while preserving long-term soil fertility and pond functionality.

3.4 Monoculture, Polyculture, and Integrated Fish Farming Systems

The organization of fish species within culture systems profoundly influences productivity, ecological efficiency, economic returns, and management complexity. **Culture system design** encompasses decisions regarding single-species versus multiple-species stocking, species combinations, size ratios, and integration with other agricultural or livestock enterprises. Understanding the principles, advantages, limitations, and management requirements of different culture systems enables selection of approaches best matched to available resources, management capability, and market demands.

Monoculture Systems

Monoculture involves culturing a single species within a production system, representing the simplest organizational approach and dominating industrial-scale aquaculture worldwide.

Principles and Characteristics

Monoculture systems focus production efforts on one species, enabling optimization of all management inputs including feeding, water quality management, health monitoring, and harvest operations for that species' specific requirements. Genetic improvement programs, specialized nutrition, and disease control protocols can be developed specifically for the target species, maximizing production efficiency.

Monoculture predominates in intensive aquaculture operations targeting species commanding premium market prices or where specialized production technologies confer competitive advantages. Examples include Atlantic salmon production in Norway and Chile, channel catfish farming in the United States, rainbow trout culture in various temperate regions, Pangasius catfish production in Vietnam, and shrimp culture globally.

Advantages of Monoculture

Management simplicity represents monoculture's primary advantage. With a single species, feeding strategies can be optimized using species-specific formulated feeds with appropriate protein levels, pellet sizes, and nutritional profiles. Growth monitoring requires tracking only one species' development, simplifying data collection and analysis. Disease diagnosis and treatment focus on pathogens affecting the target species, enabling specialized health management expertise.

Harvest efficiency improves dramatically in monoculture compared to mixed-species systems. All fish reach marketable size within a predictable timeframe, enabling complete harvest operations. Grading by size rather than species simplifies processing, and harvest timing can be optimized for market conditions and price seasonality.

Market compatibility often favors monoculture production of specific species matching consumer preferences and processing requirements. Uniform size, consistent quality, and reliable supply schedules meet wholesale and export market demands more effectively than variable, mixed-species production.

Genetic improvement progresses more rapidly in monoculture systems where selection programs focus on single species. Breeding objectives targeting growth rate, feed conversion, disease resistance, and product quality can be pursued systematically. Genetic gains of 10-20% per generation have been achieved in salmon, tilapia, carp, and catfish through selective breeding programs supporting monoculture production.

Specialized technologies including recirculating aquaculture systems, cage culture in open waters, and advanced hatchery techniques are typically developed for specific high-value species cultured in monoculture. Investment in sophisticated technology is economically justified when production focuses on species commanding premium prices.

Limitations and Challenges

Ecological inefficiency represents monoculture's fundamental limitation from a biological perspective. In monoculture systems, all individuals exploit the same food resources and ecological niche, creating direct competition and underutilizing pond productivity potential. Natural food production including benthic organisms, zooplankton not consumed by the target species, and plant materials often remains unexploited.

Resource underutilization extends to physical space, with monoculture species occupying preferred depth zones while leaving other strata underutilized. For example, monoculture of bottom-feeding catfish leaves mid-water and surface zones unoccupied, representing wasted production potential.

Disease vulnerability increases in dense monocultures where rapid disease transmission occurs within genetically similar populations. Disease outbreaks can cause catastrophic losses affecting entire production units. Genetic uniformity from selective breeding programs may reduce disease resistance diversity, potentially increasing epidemic risks.

Market risk concentrates in monoculture systems where production depends entirely on one species' market price and demand. Price fluctuations, market saturation, or shifts in consumer preferences directly impact farm profitability with no alternative species to buffer economic volatility.

Complete dependency on formulated feeds characterizes intensive monoculture, increasing production costs and vulnerability to feed price fluctuations. Feed typically represents 50-70% of variable costs in intensive monoculture systems.

Management Practices in Monoculture

Successful monoculture requires optimizing all management inputs for the target species' specific requirements.

Species selection should consider multiple factors including market demand and price stability, availability of quality seed stock, compatibility with available water sources and climate, technical knowledge and management expertise, and availability of suitable feeds. Species exhibiting rapid growth, efficient feed conversion, disease resistance, and wide market acceptance are preferred.

Stocking density must balance maximum production with maintained water quality, growth performance, and fish welfare. Optimal densities vary enormously by species and system:

- Extensive monoculture: 1,000-5,000 fish/ha
- Semi-intensive monoculture: 10,000-30,000 fish/ha
- Intensive monoculture without aeration: 30,000-80,000 fish/ha
- Intensive monoculture with aeration: 80,000-300,000 fish/ha
- Recirculating systems: 50-100 kg/m^3 or higher

Feeding programs employ species-specific formulated feeds with appropriate protein content (28-45% crude protein for most species), lipid levels (5-12%), and micronutrient supplementation. Feeding rates typically follow percentage body weight tables ranging from 5-8% for juveniles to 1-2% for adults, adjusted for temperature and observed consumption.

Disease prevention through biosecurity protocols, vaccination where available, optimal nutrition, and stress minimization forms the foundation of health management. Early detection through regular sampling, prompt diagnosis, and effective treatment protocols minimize losses when diseases occur.

Polyculture Systems

Polyculture involves simultaneous culture of multiple species within the same production unit, exploiting different ecological niches to enhance overall system productivity and resource utilization. Polyculture represents traditional aquaculture practice in Asia, refined over centuries to achieve remarkable efficiency and sustainability.

Principles and Ecological Foundation

Polyculture exploits the concept of **niche differentiation**, where different species occupy distinct ecological positions defined by feeding behavior, preferred food items, spatial distribution, and resource requirements. By stocking compatible species with complementary resource use patterns, polyculture achieves higher total productivity than monoculture of any single species.

Trophic complementarity combines species feeding at different levels within food webs. A typical freshwater polyculture might include:

- **Phytoplankton feeders** (silver carp, Hypophthalmichthys molitrix) filtering algae from the water column
- **Zooplankton feeders** (bighead carp, Hypophthalmichthys nobilis) consuming small crustaceans and larger plankton

- **Bottom feeders** (common carp, Cyprinus carpio) ingesting benthic organisms, detritus, and supplemental feed
- **Column feeders** (rohu, Labeo rohita; catla, Catla catla) consuming mid-water organisms and supplemental feed
- **Omnivores or predators** (grass carp, Ctenopharyngodon idella; snakehead, Channa spp.) controlling vegetation or rough fish

Spatial complementarity combines species occupying different vertical zones. Surface feeders, mid-water species, and bottom dwellers coexist without severe spatial competition. This three-dimensional resource exploitation increases pond carrying capacity substantially beyond monoculture levels.

Temporal complementarity can be achieved by combining species with different growth seasons or production cycles, maintaining year-round productivity in some polyculture systems.

Traditional Asian Polyculture Systems

Chinese Polyculture

The classical **Chinese polyculture** system combines six cyprinid species with complementary feeding habits, refined over millennia to achieve exceptional efficiency. The traditional combination includes:

- **Silver carp** (30-40% of total stocking): Filter-feeding on phytoplankton
- **Bighead carp** (10-15%): Filter-feeding on zooplankton
- **Grass carp** (10-20%): Herbivorous, consuming aquatic macrophytes and terrestrial vegetation
- **Common carp** (20-30%): Omnivorous bottom feeder

- **Black carp** (5-10%): Molluscivorous bottom feeder
- **Mud carp** (5-10%): Detritivorous bottom feeder

Typical stocking densities in semi-intensive Chinese polyculture range from 5,000-15,000 fingerlings per hectare, with ratios adjusted based on natural food availability and supplemental feeding levels. Annual yields commonly reach 3,000-6,000 kg/ha without supplemental aeration, and 6,000-10,000 kg/ha with aeration and intensive feeding.

Management involves organic fertilization to stimulate plankton production for filter feeders, provision of grass clippings or aquatic plants for grass carp, and supplemental feeding with formulated feeds or agricultural byproducts for omnivorous species. The system exhibits remarkable ecological efficiency, with fertilizer inputs supporting multiple species through complex food web pathways.

Indian Polyculture (Indian Major Carps)

The **Indian polyculture system** combines three indigenous carp species with distinct feeding niches:

- **Catla** (Catla catla, 30-40%): Surface feeder consuming zooplankton and insects
- **Rohu** (Labeo rohita, 30-40%): Column feeder grazing on decaying vegetation, plankton, and supplemental feed
- **Mrigal** (Cirrhinus mrigala, 20-30%): Bottom feeder consuming detritus, benthic organisms, and decaying matter

Modern intensive Indian polyculture often incorporates three additional exotic species:

- **Silver carp** (10-15%): Phytoplankton filter feeder
- **Grass carp** (5-10%): Aquatic weed control

- **Common carp** (10-15%): Bottom omnivore

This six-species **composite culture system** achieves yields of 4,000-8,000 kg/ha in semi-intensive management and 8,000-12,000 kg/ha under intensive conditions with supplemental feeding and aeration. The system has been widely adopted across South and Southeast Asia due to proven productivity and economic viability.

Stocking densities in composite culture typically range from 8,000-12,000 fingerlings/ha for semi-intensive systems and 12,000-20,000 fingerlings/ha for intensive systems. Species ratios are adjusted based on market demand, with catla and rohu often commanding premium prices.

Israeli Polyculture

Israeli-style polyculture combines common carp as the primary species with tilapia and various other species. A typical combination includes:

- **Common carp** (60-70%): Fed with pelleted feeds, representing the primary commercial species
- **Tilapia** (20-30%): Self-recruiting through natural reproduction, controlling algae and providing bonus production
- **Silver carp** (5-10%): Phytoplankton control
- **Grass carp** (2-5%): Vegetation management

This system achieves carp production of 3,000-5,000 kg/ha plus 1,000-2,000 kg/ha tilapia bonus. The presence of tilapia providing natural reproduction creates a size-graded population that exploits various food resources throughout the season while avoiding competition with larger carp.

Advantages of Polyculture

Enhanced productivity through fuller utilization of natural food resources represents polyculture's primary advantage. Total yield per hectare typically exceeds monoculture of any constituent species by 50-200%, depending on species combinations and management intensity. This productivity gain occurs without proportional increase in inputs, improving economic efficiency.

Ecological efficiency improves through complementary resource use. Natural food production including phytoplankton, zooplankton, benthic organisms, and detritus is more completely harvested through diverse species assemblages. The **ecological production efficiency** (ratio of fish production to primary production) in well-designed polycultures can reach 25-35%, compared to 10-20% in monoculture systems.

Water quality improvement occurs through biological interactions among species. Phytoplankton-feeding species control excessive algal blooms, reducing oxygen fluctuation risks. Herbivorous species manage aquatic vegetation, preventing weed problems. Bottom feeders disturb sediments, releasing nutrients and improving decomposition, though this must be balanced against turbidity risks.

Risk diversification spreads production across multiple species with different disease susceptibilities, market demands, and environmental tolerances. If one species experiences health problems or market price declines, production and income from other species provide economic buffer.

Resource recycling within integrated polyculture systems converts wastes from one species into food for others. For example, organic wastes from fish-feeding activities support bacterial and detrital food webs exploited by bottom feeders and filter feeders.

Reduced feed costs result from greater natural food utilization. Polyculture systems often achieve acceptable production with supplemental feeding of only the primary omnivorous species, while secondary species subsist primarily on natural foods. This reduces overall feed requirement per unit production compared to feeding all species in separate monocultures.

Challenges and Limitations

Management complexity increases substantially in polyculture compared to monoculture. Multiple species require different optimal conditions, and compromises must be made that may not perfectly suit any single species. Monitoring growth performance, health status, and nutritional condition requires greater expertise and effort when managing multiple species simultaneously.

Harvest difficulties arise from separating mixed species during harvest operations, particularly when market sizes differ substantially. Selective harvest of specific species is challenging, often necessitating complete pond draining. Grading mixed species requires additional labor and specialized equipment.

Market challenges occur when all species in polyculture don't command equivalent prices or demand. Low-value species may contribute substantial biomass but limited economic return. Finding buyers for multiple species simultaneously or storing different species separately for staggered marketing increases logistical complexity.

Feeding optimization becomes difficult when species have different nutritional requirements and feeding behaviors. Formulated feeds optimized for one species may not suit others' needs. Broadcast feeding may not reach all species effectively, with aggressive feeders consuming disproportionate shares.

Disease management complexity increases with multiple species potentially harboring different pathogens or exhibiting different disease susceptibilities. Diagnosing health problems requires identifying affected species and species-specific pathogens. Treatment protocols effective for one species may be toxic or ineffective for others.

Genetic improvement progresses more slowly in polyculture systems where multiple species require separate breeding programs. Resources dedicated to genetic selection must be divided among species rather than concentrated on a single target.

Predation risks emerge in some polyculture combinations. Species originally stocked at compatible sizes may develop predator-prey relationships as they grow, with larger individuals consuming smaller species. Careful species selection and size management mitigate but don't eliminate this risk.

Design Principles for Successful Polyculture

Effective polyculture design requires systematic consideration of species compatibility, ecological complementaritqy, and management objectives.

Species Compatibility Assessment

Compatible species combinations should meet multiple criteria:

- **Similar environmental requirements**: Species should tolerate similar temperature ranges, pH, salinity, and dissolved oxygen levels. Incompatibility in environmental tolerances necessitates compromises that stress some species.
- **Non-aggressive behavior**: Species should not exhibit territorial aggression or predatory behavior toward other stocked species. Aggressive species intimidate others, reducing feeding and growth.
- **Complementary feeding niches**: Species should primarily exploit different food resources, minimizing direct competition while maximizing resource utilization.
- **Compatible stocking sizes**: Initial size differences should not enable predation, typically maintaining size ratios within

3:1. Species with vastly different growth rates may develop problematic size disparities during culture.

- **Non-reproductive interference**: Species should not interbreed (producing sterile hybrids) or exhibit reproductive behaviors that disrupt others. Mouth-brooding species may refuse supplemental feed during breeding periods, reducing growth.
- **Market size synchrony**: Species should reach marketable sizes within similar timeframes to enable efficient harvest, or alternatively, early-maturing species should provide interim cash flow while later-maturing species continue growing.

Stocking Ratio Determination

Optimal species ratios depend on natural food availability, supplemental feeding intensity, and market value considerations. General principles include:

- **Natural food-based ratios**: In fertilized ponds with minimal supplemental feeding, stocking ratios should reflect estimated abundance of different food resources. If phytoplankton production is high, increase proportion of filter-feeding species. If benthic organisms are abundant, increase bottom feeders.
- **Supplemental feeding-based ratios**: When intensively feeding formulated diets, increase proportion of species accepting pelleted feeds while reducing proportions of strict natural food feeders.
- **Economic optimization**: Higher proportions should be allocated to species commanding premium market

prices, balancing economic returns against ecological complementarity.

- **Experimental optimization**: Stocking ratios should be refined through systematic trials under local conditions, as optimal combinations vary with climate, water source, management practices, and market factors.

Integrated Fish Farming Systems

Integrated fish farming combines aquaculture with other agricultural or livestock enterprises within coordinated systems where wastes or byproducts from one component provide inputs for another. Integration enhances overall farm efficiency, sustainability, and profitability through resource recycling and diversified production.

Fish-Livestock Integration

Fish-poultry integration represents the most common and productive integrated system. Chicken houses are constructed over fish ponds or adjacent to them, with manure falling or being swept directly into ponds. Chicken manure provides organic fertilization stimulating plankton production, while simultaneously reducing manure disposal problems from poultry operations.

Typical systems support 200-500 chickens per hectare of pond, producing 3,000-6,000 kg fish/ha annually with minimal supplemental feeding. Fish polycultures tolerant of high organic loading, particularly tilapia, common carp, and Chinese carps, are most suitable. Water quality management becomes critical, as excessive manure loading causes oxygen depletion and ammonia accumulation.

Fish-pig integration follows similar principles, with pigsties positioned over ponds or with slatted floors allowing manure to drop into water. Pig manure has lower nitrogen content but higher fiber than poultry manure, stimulating bacterial and zooplankton production.

Systems typically support 30-60 pigs per hectare of pond, producing 2,500-5,000 kg fish/ha.

Fish-cattle/buffalo integration is practiced where cattle sheds adjoin ponds, with manure collected and applied as pond fertilizer. Cattle manure has the lowest nutrient concentration among common livestock manures but stimulates gradual, stable plankton production. Application rates of 100-300 kg fresh weight per hectare per week maintain moderate fertility.

Advantages of fish-livestock integration:

- Manure disposal problem converted to valuable resource
- Reduced fertilizer costs for fish production
- Diversified income from multiple enterprises
- Improved overall farm resource efficiency
- Reduced environmental pollution from livestock waste

Limitations and management requirements:

- Disease transmission risks between livestock and fish require attention
- Water quality management becomes more challenging with high organic loading
- Veterinary drugs and antibiotics used in livestock may affect fish
- Market acceptance concerns if consumers perceive hygiene risks
- Regulatory restrictions on livestock-fish integration in some regions

Fish-Crop Integration

Fish-rice integration or **rice-fish culture** combines fish production with rice cultivation in paddy fields, representing an ancient practice with modern refinements. Fish are stocked in flooded rice paddies, controlling insects and weeds, providing organic fertilization through their wastes, and yielding bonus production beyond rice harvest.

Traditional rice-fish systems yield 100-500 kg fish/ha alongside normal rice production. Modern intensive systems with deeper water refuges, supplemental feeding, and carefully selected species can produce 500-1,500 kg fish/ha without reducing rice yields.

Suitable species for rice-fish culture include common carp, Nile tilapia, climbing perch, snakehead, and various local species tolerating shallow water and high temperatures. Species should not damage rice plants through feeding or physical disruption.

Fish-horticulture integration locates fish ponds within or adjacent to fruit orchards, vegetable gardens, or plantation crops. Pond water enriched with nutrients from fish wastes irrigates crops, while crop wastes and trimmings feed herbivorous fish. Fruit tree leaves falling into ponds support detrital food webs and grass carp feeding.

Fish-aquatic vegetable integration combines fish culture with production of aquatic vegetables like water spinach (Ipomoea aquatica), lotus (Nelumbo nucifera), or water chestnut (Trapa natans). Vegetables grow on pond surfaces or in shallow margins, utilizing nutrients from fish wastes while providing shade reducing temperature and algal blooms. Herbivorous fish may consume vegetable trimmings.

Integrated Recirculating Systems

Aquaponics represents modern integration combining fish culture with soilless plant cultivation, where fish waste-enriched water provides nutrients for plant growth, while plants filter and purify water for fish reuse. Aquaponic systems achieve remarkable resource efficiency with minimal water use and waste discharge.

Typical aquaponic systems maintain fish culture tanks at densities of 20-60 kg/m^3, with effluent water flowing through plant growing beds containing media (gravel, clay pellets) or nutrient film channels. Nitrifying bacteria convert fish-excreted ammonia to nitrates, which plants absorb as primary nitrogen source. Purified water returns to fish tanks, completing the cycle.

Commercial aquaponic operations produce leafy vegetables (lettuce, herbs, spinach) alongside tilapia, barramundi, or ornamental fish. Production ratios typically target 0.5-1.0 kg plants per 1 kg fish produced. Initial capital costs are high ($100-300/m²) but operational efficiency and premium pricing for "sustainably produced" products can justify investment.

Applied Implications for Culture System Selection

The choice among monoculture, polyculture, and integrated systems should reflect available resources, management capacity, market access, and production objectives. Monoculture suits intensive operations targeting high-value species with established markets and available technology. Polyculture optimizes productivity in semi-intensive systems where natural food contributes significantly to production. Integrated systems maximize overall farm efficiency through resource recycling, particularly appropriate for small-scale operations seeking diversified income and sustainability. Successful implementation requires matching system complexity to management expertise, with gradual progression from simple to complex systems as experience develops.

3.5 Aeration, Circulation, and Waste Management in Culture Systems

Intensive aquaculture systems approach or exceed natural carrying capacity limits, necessitating active environmental management through aeration, water circulation, and waste removal. Understanding the principles and technologies of these interventions enables sustainable intensification supporting high productivity while maintaining fish health and environmental quality.

Oxygen Dynamics and Aeration Requirements

Dissolved oxygen limitation represents the primary constraint on intensification in aquaculture systems. Natural oxygen production through photosynthesis and atmospheric diffusion typically supports only 300-1,000 kg fish biomass per hectare, while intensive operations may exceed 10,000-50,000 kg/ha.

Oxygen Budget Analysis

An oxygen budget quantifies all sources and sinks, determining whether supply meets demand:

DO(supply) = Photosynthesis + Atmospheric diffusion + Mechanical aeration DO(demand) = Fish respiration + Bacterial decomposition + Sediment oxygen demand + Nitrification

For intensive systems, the critical period occurs during predawn hours when photosynthetic production ceases but all consumption processes continue. Aerators must compensate for the total overnight oxygen deficit to maintain minimum DO above 4-5 mg/L.

Aeration Technologies and Performance

Paddlewheel aerators dominate pond aquaculture due to superior oxygen transfer efficiency (1.5-2.5 kg O_2/kWh), durability, and reasonable capital cost ($800-2,000 per unit). Standard 2 HP paddlewheels effectively cover 4,000-8,000 m^2 pond area. Optimal placement is 20-30 meters from shore, creating circulation patterns covering the entire pond.

Propeller-aspirator aerators provide excellent vertical mixing and destratification at 1.2-1.8 kg O_2/kWh efficiency, suitable for deep ponds prone to stratification. Cost and complexity are higher than paddlewheels.

Diffused air systems introduce compressed air through bottom diffusers, achieving good circulation but lower oxygen transfer efficiency (0.8-1.5 kg O_2/kWh). Capital costs are high due to compressor and piping requirements, but operational flexibility is excellent.

Aeration System Design

The required aeration capacity can be estimated from:

HP(required) = [Fish biomass (kg) × Feed rate (%) × 0.25] / AOT

Where AOT is the aerator oxygen transfer rate (typically 1.5-2.0 kg O_2/HP/hour for paddlewheels), and 0.25 represents the estimated oxygen consumption per kg feed applied.

For example, a pond with 50,000 kg fish fed at 2% body weight daily requires approximately 25 HP aeration capacity for emergency conditions, though only 40-60% of capacity typically operates during normal nighttime aeration.

Aeration Management Strategies

Continuous aeration operates 24 hours daily in extremely intensive systems (>100 kg/m^3 in tanks or RAS), maintaining stable oxygen levels but consuming maximum energy.

Nighttime aeration activates during evening and predawn periods when photosynthetic oxygen production ceases, typically operating from sunset to 2-3 hours after sunrise. This strategy balances oxygen maintenance with energy conservation.

Emergency aeration responds to oxygen crisis situations, typically activated when DO falls below 2.5-3.0 mg/L. All available aerator capacity operates until oxygen recovers above safe levels.

Automated aeration control using DO sensors and controllers optimizes aerator operation, reducing energy consumption by 30-50% compared to manual operation while maintaining adequate oxygen levels.

Water Circulation and Exchange

Beyond oxygen provision, water circulation distributes heat, nutrients, and dissolved substances uniformly throughout culture systems, prevents stagnant zones, and facilitates waste removal.

Circulation Patterns and Design

Effective circulation eliminates dead zones where waste accumulates and oxygen depletes. In rectangular ponds, aerator placement should create unidirectional flow patterns, with water moving along long axes and returning along shorter sides. Circular tanks inherently create rotational circulation when inlets are positioned tangentially.

Water velocity in flow-through systems or heavily circulated ponds typically ranges from 1-5 cm/second in feeding-based culture and 10-30 cm/second in raceway systems for salmonids. Excessive velocity (>30 cm/second) causes fish stress and increased energy expenditure.

Water Exchange Systems

Flow-through systems provide continuous water replacement, suitable for coldwater species requiring highly oxygenated water or high-density production. Exchange rates range from 1-10 complete water replacements per hour depending on stocking density and species requirements.

Water exchange calculation:

Q (L/min) = [Fish weight (kg) × DO consumption (mg/kg/hr)] / [60 × (DO(inlet) - DO(outlet))]

Semi-static systems employ periodic batch water exchange, typically 5-20% daily in semi-intensive ponds. Exchange timing should avoid predawn periods when pond oxygen is naturally lowest. Afternoon exchange during peak photosynthetic production minimizes oxygen disruption.

Recirculating aquaculture systems (RAS) minimize water use through extensive treatment and reuse, typically requiring makeup water of only 1-10% daily. Multiple treatment components including mechanical filtration, biofiltration, oxygenation, and disinfection enable 90-99% water recirculation.

Waste Management in Aquaculture Systems

Fish wastes include uneaten feed, feces, and metabolic products. Effective waste management prevents accumulation causing water quality deterioration and environmental impacts.

Solid Waste Characteristics and Management

Solid waste production approximates 20-30% of feed applied in intensive systems. Settleable solids concentrate in low-flow zones and pond bottom depressions, decomposing and consuming oxygen.

Settling basins located before pond inlets remove suspended solids from incoming water. Basins sized at 3-5% of pond area with retention time of 15-30 minutes allow settling of particles >30 μm.

Sludge removal systems in tanks and raceways continuously collect settled wastes. Central bottom drains in circular tanks concentrate solids for removal. Mechanical systems including drum filters, sand filters, or settling chambers remove solids before water returns to culture units.

Pond sediment management requires periodic removal when accumulation exceeds 15-20 cm. Mechanical dredging or manual excavation removes nutrient-rich sediments for agricultural use or disposal.

Dissolved Waste Management

Nitrogenous waste management has been addressed in water quality sections. Additional strategies include:

Wetland treatment systems located downstream of aquaculture effluents utilize emergent plants (reeds, cattails) to assimilate nitrogen and phosphorus. Wetlands sized at 5-15% of pond area can remove 30-60% of dissolved nutrients from effluent.

Algal turf scrubbers use shallow channels with attached algae to remove nutrients from recirculating water. Algae are periodically harvested, permanently removing nutrients from the system.

Denitrification reactors in RAS provide anoxic zones with carbon sources where denitrifying bacteria convert nitrate to nitrogen gas, achieving 50-80% nitrate removal. Reactor volume typically equals 1-3% of total system volume.

Environmental Regulations and Best Management Practices

Aquaculture operations increasingly face environmental regulations limiting effluent discharge, requiring waste treatment or adoption of best management practices (BMPs).

Effluent Quality Standards typically limit total suspended solids (TSS <30-50 mg/L), biochemical oxygen demand (BOD <30-50 mg/L), total nitrogen (<5-10 mg/L), and total phosphorus (<1-2 mg/L) in discharged water. Meeting these standards requires treatment or management modifications.

Best Management Practices include:

- Optimizing feeding to minimize waste (FCR <1.5:1)
- Using high-quality feeds with digestible ingredients
- Settling ponds or wetlands for effluent treatment

- Timing water exchange to minimize environmental impact
- Periodic sediment removal preventing excessive accumulation
- Adopting recirculating or integrated systems reducing discharge

Applied Implications for Intensive Culture Management

Successful intensification beyond natural carrying capacity requires investment in aeration infrastructure, electrical power supply, and backup systems preventing catastrophic oxygen depletion. Reliable emergency power generation protects against grid failures that could cause complete crop loss. Waste management must be integrated into system design rather than addressed as an afterthought, with facilities for solid waste removal and treatment of liquid effluents. Sustainable intensification balances production goals with environmental stewardship, maintaining long-term viability through responsible resource management.

3.6 Maintenance and Record Keeping in Fish Ponds

Systematic maintenance and comprehensive record keeping represent essential but often neglected management practices determining long-term productivity and profitability of aquaculture operations.

Infrastructure Maintenance

Embankment Maintenance

Regular inspection identifies problems before they cause catastrophic failures. Weekly during culture periods and after heavy rainfall, embankments should be examined for seepage, cracks, slumping, animal burrows, and erosion.

- **Seepage** through embankments creates wet areas on outer slopes, indicating water loss and potential structural weakening. Minor seepage often self-heals as fine particles clog voids, but persistent or increasing seepage requires repair. Affected sections should be drained, excavated to remove compromised material, and reconstructed with proper compaction.
- **Animal burrows** from crabs, crayfish, rodents, or snakes create channels allowing rapid water flow that can erode into catastrophic breaches. Burrows should be located, animals removed, and holes filled with clay and compacted thoroughly. Persistent burrowing problems may require installation of fencing or screens preventing access.
- **Erosion** from wave action, rainfall, or overflow concentrates on inner slopes and crests. Establishing grass cover, applying riprap, or reducing water levels during high wind periods prevents progressive erosion. Eroded areas should be refilled and compacted during maintenance periods.
- **Cracks** in embankments indicate drying, settlement, or structural movement. Shallow surface cracks (<3 cm deep) typically pose minimal risk but deeper cracks require filling with clay and watering to prevent erosion enlargement. Horizontal cracks parallel to embankment slopes suggest deep slumping requiring major reconstruction.

Inlet and Outlet Structures

Monthly inspection of pipes, screens, valves, and concrete structures identifies corrosion, clogging, or damage. Screens should be cleaned weekly during operation to maintain water flow. Wooden components in monk structures require annual inspection for rot and replacement every 3-5 years. Concrete structures should be examined for cracks allowing seepage around pipes.

Anti-seepage collars on pipes should be inspected during draining for detachment or rust. Failed collars must be repaired or replaced before refilling to prevent piping failures that could breach embankments.

Equipment Maintenance

Aerator Maintenance

Paddlewheel aerators require weekly greasing of bearings, monthly oil changes in gearboxes, and quarterly inspection of shafts, paddles, and mounting hardware. Worn or damaged paddles should be replaced immediately as imbalanced units cause excessive wear. Electrical connections must remain dry and properly insulated.

Annual maintenance includes complete disassembly, bearing replacement, seal inspection, gear oil change, and repainting of metal components. Properly maintained paddlewheels operate effectively for 5-8 years before major component replacement becomes necessary.

Pump Maintenance

Water pumps require daily inspection during operation, checking for unusual noise, vibration, or heating. Impellers should be cleaned monthly to remove debris. Motor bearings need lubrication following manufacturer specifications. Intake screens require cleaning to maintain flow rates.

Seasonal maintenance includes impeller inspection for wear, seal replacement if leaking, electrical connection checking, and motor insulation testing. Backup pumps should be tested monthly to ensure readiness for emergencies.

Water Quality Monitoring

Systematic water quality monitoring has been detailed previously but warrants emphasis in maintenance context. Daily dissolved oxygen measurements during critical predawn periods, weekly pH and temperature checks, and monthly comprehensive testing provide data enabling proactive management preventing crises rather than reactive emergency responses.

Monitoring Equipment Maintenance

DO meters require weekly calibration using air-saturated water or oxygen-saturated solution. Probe membranes should be replaced monthly or when response becomes slow. pH meters need two-point calibration before each use and probe storage in appropriate buffer solution. Annual laboratory calibration verification ensures measurement accuracy.

Record Keeping Systems

Comprehensive records document all management activities, production data, and economic information, enabling analysis of performance, identification of problems, and planning of future improvements.

Essential Production Records

Pond records document stocking dates and quantities, species and sizes stocked, source of fingerlings, feeding records (dates, quantities, feed types), growth sampling data (dates, sample sizes, average weights), water quality measurements, disease occurrences and treatments, partial harvests, and final harvest data.

Individual pond record sheets or digital entries in farm management software should be completed daily during culture periods. Data integrity is critical—estimates or guesses undermine record value.

Feeding Records

Daily feeding records document:

- Date and time of feeding
- Amount of feed applied by type and size
- Weather conditions affecting feeding
- Fish feeding response observations
- Estimated feed consumption

These records enable calculation of cumulative feed applied, current feed conversion ratio estimates, and feeding strategy adjustments. Correlation of feeding records with growth data identifies optimal feeding levels maximizing efficiency.

Growth Monitoring Records

Monthly growth sampling provides data tracking fish development:

- Sampling date
- Number of fish sampled
- Individual or average weights
- Length measurements for condition factor assessment
- Calculated growth rates since previous sampling
- Estimated total pond biomass

- Projected harvest dates

Growth data plotted over time reveals whether fish are achieving expected performance. Deviation from expected growth curves indicates problems requiring investigation—possible underfeeding, disease, or water quality issues.

Water Quality Records

Systematic recording of water quality parameters creates historical database enabling pattern recognition:

- Date and time of measurement
- Location in pond (surface, mid-depth, bottom if relevant)
- Temperature (°C)
- Dissolved oxygen (mg/L)
- pH
- Secchi depth (cm)
- Other parameters tested

Graphing data over culture periods reveals seasonal patterns, correlations between parameters, and relationships to management activities. For example, declining morning DO over several weeks despite unchanged feeding rates might indicate accumulating sediment oxygen demand requiring intervention.

Disease and Mortality Records

Health observations and mortality events require careful documentation:

- Date and time first observed

- Number of affected fish
- Clinical signs and behavioral changes
- Dead fish counts and estimated mortality
- Diagnostic tests performed and results
- Treatments applied (drug, dose, duration)
- Treatment outcomes and resolution dates

Disease records enable pattern analysis identifying predisposing factors, effectiveness of treatments, and frequency of specific conditions. This information guides preventive management and improves response to future outbreaks.

Economic Records

Financial records track costs and revenues enabling profitability analysis:

Input costs:

- Fingerling purchase costs
- Feed costs by quantity and type
- Fertilizer and chemical costs
- Labor costs
- Fuel and electricity costs
- Equipment purchase and repairs
- Interest on loans

Output revenues:

- Harvest quantities by species and size

- Sale prices received
- Total revenue per pond and farm
- Buyer information

Economic analysis calculates:

- Cost per kg fish produced
- Revenue per kg harvested
- Profit per pond and per hectare
- Return on investment
- Feed cost as percentage of total cost
- Labor efficiency (kg produced per labor hour)

Data Analysis and Management Decisions

Records transform from historical documentation to management tools through systematic analysis. Key analyses include:

Feed conversion ratio tracking = Total feed applied / Net fish production, calculated monthly and cumulatively. FCR increasing over time suggests overfeeding, deteriorating feed quality, or health problems.

Survival rate = (Final number / Initial number) × 100, calculated at harvest. Comparing survival across ponds and production cycles identifies problem ponds or seasonal patterns.

Specific growth rate = $[(\ln W_2 - \ln W_1) / \text{Days}] \times 100$, where W_1 and W_2 are initial and final weights. SGR declining below expected rates indicates growth problems requiring investigation.

Partial productivity analysis examines relationships between specific inputs and outputs:

- kg fish produced per kg feed applied

- kg fish per unit labor hour
- kg fish per HP aeration capacity
- Revenue per unit of limiting resource

Benchmarking compares farm performance to industry standards or best-performing operations, identifying gaps and improvement opportunities. Yield of 3,000 kg/ha when regional average is 6,000 kg/ha signals substantial improvement potential through better management.

Digital Record Systems and Software

Traditional paper-based records increasingly transition to digital systems offering advantages of data backup, rapid calculations, graphical visualization, and integration across farm operations.

Spreadsheet-based systems using software like Microsoft Excel or Google Sheets provide customizable, low-cost solutions suitable for small to medium operations. Templates can track daily activities, automatically calculate growth metrics, and generate graphs visualizing trends.

Specialized aquaculture management software offers comprehensive functionality including pond tracking, feeding management, growth prediction models, inventory management, financial analysis, and regulatory reporting. Commercial packages cost $500-5,000 annually depending on scale but provide sophisticated analytical capabilities.

Mobile applications enable real-time data entry during pond inspections using smartphones or tablets, with data automatically syncing to central databases. Digital photography documenting infrastructure conditions, fish health, and water appearance enhances record quality.

Data Security and Backup

Digital records require protection against loss through hardware failures, software corruption, or cyber threats. Essential practices include:

- Daily or weekly backup to separate physical locations
- Cloud-based storage with automatic synchronization
- Version control preventing accidental overwriting
- Password protection and access controls
- Annual archive transfer to permanent storage media

Staff Training and Standard Operating Procedures

Consistent implementation of maintenance and record keeping requires trained personnel following documented procedures.

Standard Operating Procedures (SOPs) detail specific protocols for routine activities:

- Aerator startup and shutdown procedures
- Feeding protocols specifying rates, timing, and adjustments
- Water quality testing methods and equipment calibration
- Disease response protocols including diagnostics and treatment
- Emergency procedures for oxygen depletion, disease outbreaks, or equipment failures

SOPs should be written clearly with step-by-step instructions, illustrated with photos or diagrams, translated into relevant languages, and readily accessible to workers. Annual review updates procedures based on accumulated experience.

Staff training programs ensure personnel understand their responsibilities and possess required skills. New employee orientation should cover farm layout, pond identification systems, equipment operation, safety procedures, and record keeping requirements. Ongoing training addresses new techniques, equipment, or procedures as they are adopted.

Applied Implications for Farm Management

Maintenance and record keeping represent investments in long-term sustainability and continuous improvement. Systematic infrastructure inspection and repair prevents catastrophic failures that could destroy entire crops. Equipment maintenance ensures reliability during critical periods when failures could cause massive losses. Comprehensive records enable data-driven decision making, identifying successful practices worth replicating and problematic patterns requiring correction. Farms implementing rigorous maintenance and record keeping consistently outperform those relying on intuition and reactive crisis management, validating these practices as essential components of professional aquaculture operations.

CHAPTER 4

NUTRITION, FEEDING, AND GROWTH MANAGEMENT

Key Outcomes

1. Optimal nutrition is essential for maximizing fish growth, health, feed efficiency, and profitability in aquaculture.
2. Proteins are the most critical and costly macronutrient, forming 28–50% of formulated feeds depending on species and life stage.
3. Fish utilize protein preferentially for both energy and tissue synthesis, unlike terrestrial animals that rely more on carbohydrates.
4. Species-specific and stage-specific protein requirements must be met to ensure efficient growth and immune function.
5. Essential amino acids (EAAs) must be supplied through diet, as deficiencies can limit growth even if total protein is adequate.
6. Balanced amino acid profiles and digestible protein sources are key to cost-effective and biologically efficient feed formulation.

4.1 Nutritional Requirements of Cultivable Fishes

Optimal nutrition represents the foundation of successful aquaculture, directly determining growth rates, feed conversion efficiency, health status, and ultimately economic profitability. Fish require specific nutrients in appropriate quantities and ratios to support maintenance metabolism, growth, reproduction, and immune function. Understanding species-specific nutritional requirements enables formulation of cost-effective diets maximizing production efficiency.

Macronutrients: Proteins and Amino Acids

Protein represents the most expensive and critical nutrient in fish diets, typically comprising 28-50% of formulated feeds depending on species and life stage. Unlike terrestrial animals that can synthesize energy from carbohydrates, many fish species preferentially utilize protein for both energy and tissue synthesis, elevating protein requirements substantially.

Dietary protein requirements vary by species:

- Carnivorous species (salmon, trout, grouper): 40-50% crude protein
- Omnivorous species (carp, tilapia, catfish): 28-35% crude protein
- Herbivorous species (grass carp): 25-30% crude protein
- Larval stages: 45-55% crude protein (all species)

Protein requirements decline with increasing fish size, as maintenance metabolism becomes proportionally larger relative to growth. Juveniles may require 40-45% protein while adults of the same species grow adequately on 30-35% protein.

Essential amino acids represent the building blocks of proteins that fish cannot synthesize and must obtain from diet. Ten amino acids are considered essential for fish: arginine, histidine, isoleucine, leucine, lysine, methionine, phenylalanine, threonine, tryptophan, and valine. Cystine and tyrosine are semi-essential, sparing methionine and phenylalanine requirements respectively.

Amino acid balance in dietary protein profoundly affects protein utilization efficiency. Deficiency in any single essential amino acid limits overall protein synthesis, with excess amino acids deaminated and used for energy or excreted, wasting expensive protein sources. High-quality protein sources providing amino acid profiles matching fish requirements enable superior growth performance at lower total protein levels compared to imbalanced proteins.

Animal protein sources (fish meal, meat meal, blood meal) generally provide superior amino acid balance for carnivorous species, while plant proteins (soybean meal, canola meal) suit omnivorous species better. Complementary protein combinations mixing animal and plant sources often achieve better amino acid balance than single protein sources.

Energy: Lipids and Carbohydrates

Lipids (fats and oils) provide concentrated energy (approximately 9 kcal/g compared to 4 kcal/g for proteins or carbohydrates), supply essential fatty acids, facilitate vitamin absorption, and improve feed palatability. Optimal dietary lipid levels vary by species:

- Salmonids and marine species: 15-25% lipid
- Warm-water omnivores: 5-12% lipid
- Herbivorous species: 3-6% lipid

Essential fatty acids (EFA) including omega-3 and omega-6 polyunsaturated fatty acids (PUFA) must be provided in diets. Freshwater fish generally require both linoleic acid (18:2n-6) and linolenic acid (18:3n-3), while marine fish specifically require long-chain omega-3 fatty acids including eicosapentaenoic acid (EPA, 20:5n-3) and docosahexaenoic acid (DHA, 22:6n-3).

EFA deficiency causes reduced growth, fin erosion, fatty liver, increased disease susceptibility, and impaired reproduction. Typical EFA requirements range from 0.5-2.0% of diet depending on species.

Carbohydrates provide energy at lower cost than proteins or lipids but are utilized with variable efficiency by fish. Herbivorous and omnivorous warm-water species tolerate high dietary carbohydrate levels (30-40%) with good utilization. Carnivorous species exhibit limited carbohydrate digestion capacity and poor glucose regulation, typically limiting dietary carbohydrates to <20% of diet.

Simple carbohydrates (sugars) are digested more readily than complex carbohydrates (starches), and cooking or extrusion improves starch digestibility substantially. Raw starches exhibit digestibility of 40-60% while properly processed starches reach 80-95% digestibility.

Vitamins and Minerals

Vitamins function as coenzymes in metabolic processes and are required in small quantities (mg or µg per kg diet). Fish require the same vitamins as terrestrial animals including fat-soluble vitamins (A, D, E, K) and water-soluble vitamins (B-complex, vitamin C).

Vitamin C (ascorbic acid) requires special attention in fish nutrition, as fish cannot synthesize it and it degrades rapidly in feeds exposed to heat, light, or moisture. Vitamin C deficiency causes scoliosis (spinal deformation), impaired wound healing, hemorrhaging, and reduced disease resistance. Dietary requirements range from 50-200 mg/kg diet depending on species. Stabilized forms (ascorbyl phosphate, ascorbyl sulfate) resist degradation better than pure ascorbic acid.

Vitamin E serves as an antioxidant protecting cell membranes from oxidative damage, particularly important in diets containing high levels of polyunsaturated lipids. Requirements range from 50-100 mg/kg diet, increasing with dietary lipid content.

Minerals serve structural roles (calcium and phosphorus in bones) and functional roles (iron in hemoglobin, iodine in thyroid hormones). Fish absorb some minerals directly from water, particularly calcium and sodium, reducing dietary requirements relative to terrestrial animals.

Phosphorus represents the most important dietary mineral, as water contains insufficient phosphorus for maximal growth. Requirements range from 0.4-0.9% available phosphorus in diet depending on species and life stage. Phosphorus deficiency causes reduced growth, skeletal deformities, and poor bone mineralization.

Other essential minerals include magnesium, potassium, iron, zinc, copper, manganese, selenium, iodine, and cobalt. **Premixes** containing vitamins and trace minerals at appropriate levels are incorporated into formulated feeds at 0.5-2.0% inclusion rates.

Digestibility and Nutrient Availability

Apparent digestibility measures the proportion of nutrients absorbed from the digestive tract:

AD (%) = [(Nutrient intake - Fecal nutrient) / Nutrient intake] × 100

Digestibility varies with ingredient quality, processing methods, and fish species. Animal proteins generally exhibit higher digestibility (85-95%) than plant proteins (70-85%). Lipids are highly digestible (85-95%) across sources. Carbohydrate digestibility ranges from 40-95% depending on source and processing.

Anti-nutritional factors in some ingredients reduce nutrient availability. Trypsin inhibitors in raw soybeans inhibit protein digestion, phytate binds phosphorus making it unavailable, and gossypol in cottonseed meal exhibits toxicity at high levels. Processing methods including heating, fermentation, or enzyme treatment reduce anti-nutritional factors.

Life Stage and Environmental Effects on Requirements

- **Larvae and fry** have highest protein and energy requirements (45-55% protein, 10-15% lipid) supporting rapid growth during this critical period. Diets must contain small particles (<500 μm initially) with high digestibility.
- **Juveniles** maintain elevated requirements (38-45% protein) but can utilize slightly larger particles and less expensive ingredients.

- **Adults** have lowest requirements (28-35% protein for most species) as maintenance metabolism dominates and growth slows.
- **Broodstock** require specialized nutrition including enhanced levels of vitamins (especially vitamin E and C), essential fatty acids (EPA and DHA), and astaxanthin or other carotenoids improving egg quality and larval survival.
- **Temperature** affects metabolic rate and thus nutrient requirements. At temperatures near optimal range, protein requirements may be slightly elevated to support faster growth. At suboptimal temperatures, energy requirements for maintenance increase while growth slows.
- **Salinity** affects osmoregulatory energy expenditure, with fish in suboptimal salinities requiring additional energy for osmotic regulation.

Applied Implications for Feeding Management

Understanding nutritional requirements enables selection or formulation of diets optimized for specific production situations. Using feeds with excessive protein levels wastes money and increases nitrogen pollution, while inadequate protein limits growth. Matching diet specifications to species, size, and environmental conditions maximizes growth efficiency and economic returns while minimizing environmental impacts.

4.2 Types of Fish Feed – Natural and Artificial

Fish feeds are classified into natural foods produced within culture systems and artificial feeds provided by farmers. Most production systems utilize combinations of both feed types, balancing cost, availability, and nutritional completeness.

Natural Foods

Natural foods develop spontaneously in ponds through primary and secondary productivity, providing complete or supplemental nutrition depending on species and system management.

Phytoplankton comprises microscopic algae including green algae (Chlorophyceae), diatoms (Bacillariophyceae), blue-green algae (Cyanophyceae), and others. Phytoplankton serves as direct food for filter-feeding species (silver carp, bighead carp, tilapia fry) and forms the base of aquatic food webs supporting higher trophic levels.

Phytoplankton production through pond fertilization yields 50-200 kg dry matter per hectare per day in productive systems, supporting 1,000-3,000 kg fish/ha annually in extensively managed ponds.

Zooplankton including rotifers, cladocerans, and copepods are consumed by larval and juvenile fish of most species. Zooplankton production lags phytoplankton blooms by 1-2 weeks as populations develop. Densities of 20-100 organisms per liter provide excellent natural food for small fish.

Benthic organisms including chironomid larvae, tubificid worms, mollusks, and crustaceans are exploited by bottom-feeding species (common carp, catfish). Benthic productivity in fertile ponds reaches 20-80 kg dry weight per hectare per day.

Insects including terrestrial insects falling into ponds and emerging aquatic insects provide protein-rich food for surface-feeding species, contributing 10-50 kg/ha/year in tropical ponds.

Macrophytes (aquatic plants) are consumed by herbivorous species like grass carp. Submerged plants (Hydrilla, Najas), floating plants (duckweed, water hyacinth), and emergent plants (water spinach) provide abundant biomass in nutrient-rich waters.

Supplementary Feeds

- **Supplementary feeds** complement natural food production with locally available agricultural or animal byproducts. Common materials include:
- **Rice bran**: 12-14% protein, 10-15% lipid, widely available and economical. Applied at 1-3% biomass daily.
- **Wheat bran**: 15-17% protein, lower cost than rice bran with similar nutritional value.
- **Oil seed cakes** (groundnut, cottonseed, soybean, coconut cakes): 20-40% protein depending on source, providing protein supplementation at 1-2% biomass daily.
- **Brewers grains and distillers grains**: 20-27% protein, byproducts from brewing and distilling industries.
- **Silkworm pupae**: 50-55% protein, 25-30% lipid, excellent nutrition but limited availability and high cost.

Supplementary feeding combines 2-4 ingredients in simple mixtures, typically providing 25-35% crude protein and supporting production of 2,000-4,000 kg/ha in semi-intensive systems. Feed conversion ratios range from 3:1 to 8:1, reflecting incomplete nutrition and variable digestibility.

Complete Formulated Feeds

Complete feeds provide all required nutrients in balanced proportions, enabling intensive production independent of natural food. Modern aquaculture feeds are scientifically formulated, manufactured using sophisticated processing technologies, and quality-controlled for consistency.

- **Feed Formulation** combines multiple ingredients meeting nutritional specifications at minimum cost:
- **Protein sources**: Fish meal (60-72% protein), soybean meal (44-48% protein), canola meal (36-38% protein), meat and bone meal (50-55% protein), poultry byproduct meal (55-60% protein), corn gluten meal (60-65% protein).
- **Energy sources**: Fish oil (100% lipid), vegetable oils (soy, canola, palm oil), wheat flour (10-12% protein, 60-70% carbohydrate), rice broken (8-10% protein, 70-75% carbohydrate).
- **Binders**: Wheat gluten, bentonite clay, lignin sulfonate improve pellet stability in water.
- **Additives**: Vitamin and mineral premixes, attractants (fish protein hydrolysate, squid meal), antioxidants (ethoxyquin, BHT), pigments (astaxanthin, canthaxanthin).
- **Least-cost formulation** using linear programming software identifies ingredient combinations meeting nutritional constraints at minimum cost, considering current ingredient prices and availability.

Manufacturing Processes

Steam pelleting: Ingredients are ground, mixed, steam-conditioned at 80-90°C, and forced through dies forming pellets. Pellets are dried to 10-12% moisture and cooled. Steam pelleting produces dense, sinking pellets suitable for bottom feeders.

Extrusion: Ingredients are mixed with moisture (20-30%), heated under pressure (110-150°C), and forced through dies. Rapid pressure release causes expansion, creating floating pellets with reduced density. Extrusion improves starch gelatinization and digestibility, produces floating or slow-sinking feeds, and enables incorporation of high lipid levels (15-25%). Modern extrusion dominates production of high-performance salmonid and marine fish feeds.

Crumbling and sizing: Larger pellets are crushed and sieved to produce appropriate sizes for different fish stages, ranging from <0.5 mm crumbles for larvae to 9-12 mm pellets for large fish.

Physical Feed Characteristics

Pellet size must match fish mouth dimensions. General guidelines suggest pellet diameter = 1/20 to 1/40 of fish length, or approximately 1/3 to 1/2 of mouth width.

Water stability determines how long pellets retain integrity before disintegrating. Quality feeds maintain structure for 30-60 minutes in water, minimizing waste and water quality deterioration.

Floatability affects feeding behavior and feed management. Floating feeds allow observation of feeding activity and enable adjustments minimizing waste. Slow-sinking feeds suit mid-water feeders. Fast-sinking feeds are consumed by bottom feeders.

Feed Quality and Storage

Feed quality depends on ingredient freshness, manufacturing precision, and proper storage. Key quality indicators include:

Nutrient content matching label specifications (verified through chemical analysis) **Pellet durability** (>95% intact pellets, minimal fines) **Absence of mold, rancidity, or contamination Appropriate moisture content** (<12%)

Storage requirements: Feeds should be stored in dry, cool, ventilated facilities protected from moisture, pests, and direct sunlight. Bags should be stacked on pallets allowing air circulation. **Shelf life** of standard feeds is 3-4 months; high-lipid feeds (>15% fat) should be used within 2 months to prevent lipid oxidation.

Applied Implications: Natural food exploitation through fertilization reduces feeding costs in semi-intensive systems. Supplementary feeding balances cost and performance for mid-level production. Complete formulated feeds enable maximum intensification and growth rates but require careful cost-benefit analysis, as feed represents 50-70% of variable production costs.

4.3 Feed Formulation and Supplementation

Commercial feed formulation employs systematic approaches balancing nutritional adequacy with economic feasibility. Small-scale operations may formulate farm-made feeds, while large operations typically purchase commercially manufactured feeds.

Formulation Principles

Linear programming identifies ingredient combinations meeting nutritional constraints at minimum cost. Constraints include minimum and maximum limits for crude protein, digestible energy, essential amino acids, fatty acids, phosphorus, fiber, and individual ingredients.

Example formulation for 32% protein catfish feed:

Ingredients (kg/100 kg feed):

- Soybean meal (48% CP): 45 kg
- Fish meal (60% CP): 8 kg
- Rice bran (12% CP): 25 kg
- Fish oil: 3 kg
- Wheat flour (binder): 15 kg
- Vitamin/mineral premix: 1 kg

- Antioxidant: 0.1 kg

This combination provides approximately 32% crude protein, 6% lipid, 2,850 kcal/kg digestible energy, meeting catfish requirements at reasonable cost.

Small-Scale Feed Manufacturing

On-farm feed production suits small to medium operations where commercial feeds are unavailable or excessively expensive. Basic requirements include:

Equipment: Grinder/mill for ingredient processing, mixer for homogenization, pelleting machine or simple hand pelletizers, drying facilities.

Process: Ingredients are ground to appropriate particle size, weighed according to formulation, thoroughly mixed, moisture added (20-30%), mixture pelleted, pellets sun-dried or mechanically dried to <12% moisture, and stored in sealed containers.

Quality control: Regular analysis of ingredient composition, consistent weighing and mixing, proper moisture management, and systematic testing of feed performance guide quality maintenance.

Limitations: Labor-intensive production, inconsistent pellet quality, limited storage stability, and difficulty incorporating vitamin/mineral premixes uniformly constrain on-farm manufacturing. Economic analysis must compare production costs including labor with commercial feed prices.

Feed Additives and Functional Ingredients

Probiotics: Beneficial bacteria (Bacillus spp., Lactobacillus spp.) added at 10^6-10^8 CFU/kg feed reportedly improve digestion, immunity, and disease resistance. Scientific evidence supporting probiotic efficacy varies.

Prebiotics: Non-digestible carbohydrates (mannanoligosaccharides, fructooligosaccharides) at 0.1-0.5% of diet stimulate beneficial gut bacteria proliferation.

Immunostimulants: β-glucans, peptidoglycans, and other compounds at 0.1-1.0% inclusion reportedly enhance immune responses, though efficacy depends on species, dose, and duration.

Enzymes: Phytase (500-1,000 FTU/kg) improves phosphorus availability from plant ingredients. Proteases and carbohydrases enhance nutrient digestibility, enabling increased plant protein inclusion.

Attractants: Amino acid mixtures, betaine, and fish protein hydrolysates at 0.5-2.0% improve feed palatability and consumption, particularly important for carnivorous species transitioning to plant-based diets.

Applied Implications: Effective formulation balances nutritional completeness with ingredient costs. Regular reformulation responding to ingredient price changes optimizes economics. Functional ingredients may provide benefits but require cost-effectiveness validation through production trials.

4.4 Feeding Techniques and Ration Scheduling

Feeding management directly determines growth performance, feed conversion efficiency, and production profitability. Optimal strategies consider species biology, environmental conditions, and economic factors.

Feeding Rate Determination

Feeding rates are typically expressed as percentage of body weight per day (% BW/day), declining as fish grow:

Tilapia example:

- 1-5 g fish: 6-8% BW/day
- 5-20 g: 5-6% BW/day
- 20-100 g: 3-5% BW/day
- 100-300 g: 2-3% BW/day
- 300 g: 1.5-2% BW/day

Temperature significantly affects feeding rates. At suboptimal temperatures, rates should be reduced by 25-50%. Feeding ceases when temperatures fall below species-specific minimums (typically 12-18°C depending on species).

Satiation feeding allows fish to consume ad libitum, maximizing growth but risking overfeeding and waste. **Restricted feeding** at 80-90% of satiation reduces feed costs slightly with minimal growth compromise.

Biomass estimation enables accurate feeding rate calculation:

Current biomass = Initial number × Average weight × Survival rate

Average weight is determined through monthly growth sampling of 50-100 fish. With feed conversion ratio estimates, biomass can be calculated as:

Biomass = Previous biomass + (Cumulative feed / Expected FCR)

Feeding Frequency and Timing

Feeding frequency affects growth and feed utilization:

- Larvae and fry: 4-8 times daily
- Fingerlings: 3-4 times daily
- Juveniles and adults: 1-2 times daily

Multiple daily feedings improve growth in young fish with limited gastric capacity. Mature fish grow adequately with once-daily feeding, reducing labor costs.

Feeding timing should avoid periods of low dissolved oxygen (pre-dawn hours) and extremely high temperatures (midday in summer). Optimal times are typically mid-morning and late afternoon when oxygen levels are acceptable and fish are actively feeding.

Feeding Methods

Hand feeding allows observation of feeding response and minimizes waste through adjusting quantities to consumption. Labor-intensive but suitable for small operations and high-value species.

Demand feeders dispense feed when fish activate trigger mechanisms, enabling self-feeding. Reduces labor but may cause overfeeding if not properly adjusted. Suitable for salmonids and catfish in tanks or cages.

Automatic feeders dispense predetermined feed quantities on programmed schedules. Timers or computer controllers regulate feeding frequency and amounts. Essential for large operations but lack ability to adjust for actual fish appetite.

Feeding response observation: Vigorous surface activity and rapid feed consumption indicate good appetite. Slow consumption or uneaten feed suggests overfeeding, health problems, or environmental stress requiring feeding rate reduction.

Environmental and Management Adjustments

Feeding should be reduced or suspended when:

- Dissolved oxygen <4 mg/L
- Temperature outside optimal range

- Severe weather conditions (storms, heavy rainfall)
- After handling stress (grading, transport, treatment)
- During disease outbreaks
- Water exchange or pond treatment operations

Appetite loss signals environmental or health problems requiring investigation before resuming full feeding rates.

Applied Implications

Optimal feeding balances maximum growth with economic efficiency. Underfeeding limits production potential, while overfeeding wastes expensive feed, deteriorates water quality, and increases disease risk. Systematic adjustment of feeding rates based on growth monitoring, environmental conditions, and feeding response optimizes performance.

4.5 Feed Conversion Efficiency and Growth Monitoring

Feed conversion ratio (FCR) represents the primary metric of feeding efficiency:

FCR = Total feed applied (kg) / Net fish production (kg)

Lower FCR indicates better efficiency. Modern intensive systems achieve FCR of 0.9-1.3:1 for salmonids, 1.2-1.8:1 for catfish and tilapia, and 1.5-2.5:1 for carps. Semi-intensive polyculture FCR ranges from 2:1 to 4:1.

Economic feed conversion = Feed cost / Value of fish produced, provides economic perspective beyond biological efficiency.

Growth Performance Metrics

Specific growth rate (SGR): SGR (%/day) = $[(\ln W_2 - \ln W_1) / \text{Days}] \times 100$

Where W_1 and W_2 are initial and final weights. SGR typically ranges from 1.5-3.5%/day depending on species, size, and conditions.

Condition factor (K): $K = (\text{Weight in g} / \text{Length}^3 \text{ in cm}) \times 100$

Values of 1.0-2.0 indicate healthy, well-fed fish. Declining K suggests inadequate nutrition or health problems.

Protein efficiency ratio (PER): PER = Fish weight gain / Protein consumed

PER of 1.5-2.5 indicates efficient protein utilization.

Growth Monitoring Procedures

- **Monthly sampling**: Sample 50-100 fish randomly, weigh individually or in groups, measure total length of subsamples, calculate average weight and condition factor, estimate total pond biomass, and adjust feeding rates accordingly.
- **Data analysis**: Plot growth curves comparing actual performance to expected growth models. Deviations signal problems requiring investigation. Calculate cumulative FCR and project harvest dates.

Factors Affecting Growth and FCR

- **Genetic factors**: Selectively bred strains grow 15-30% faster than wild populations. Hybrid vigor in crosses (tilapia, carp) improves performance.

- **Environmental factors**: Temperature, dissolved oxygen, pH, and ammonia within optimal ranges maximize growth. Suboptimal conditions increase FCR through elevated maintenance metabolism and reduced feed intake.
- **Feed quality**: High-quality ingredients with good digestibility, balanced amino acids, and fresh lipids produce better FCR than poor-quality feeds.
- **Feeding management**: Appropriate rates and timing optimize efficiency. Overfeeding increases FCR; underfeeding limits growth.
- **Fish size**: Larger fish exhibit slower growth rates and higher FCR than juveniles due to proportionally greater maintenance requirements.
- **Stocking density**: Moderate densities optimize growth. Excessive crowding increases stress, aggression, and competition, elevating FCR.
- **Health status**: Parasites, bacterial infections, or physiological stress impair growth and feed efficiency.

Applied Implications

Regular growth monitoring enables early detection of problems and timely management adjustments. Systematic FCR calculation identifies inefficiencies in feeding or environmental management. Targeting FCR improvements through optimized nutrition, feeding practices, and health management directly enhances profitability in feed-intensive aquaculture systems.

4.6 Advances in Nutritional Biotechnology in Pisciculture

Modern nutritional science increasingly employs biotechnology tools to enhance feed efficiency, sustainability, and fish performance.

Plant Protein Optimization

Anti-nutritional factor removal: Genetic modification or selective breeding produces soybean varieties with reduced trypsin inhibitors, phytate, and oligosaccharides, improving digestibility and amino acid availability.

Enzyme supplementation: Exogenous phytase added to feeds liberates phosphorus from phytate in plant ingredients, enabling higher plant protein inclusion while reducing phosphorus pollution. Phytase supplementation allows 20-40% reduction in supplemental phosphorus while maintaining performance.

Novel Protein Sources

Insect meals: Black soldier fly larvae (Hermetia illucens), mealworms (Tenebrio molitor), and crickets provide 40-60% protein with favorable amino acid profiles. Insect production on organic waste streams creates circular economy opportunities. Current high costs limit commercial adoption, but technological improvements promise cost competitiveness.

Single-cell proteins: Bacterial protein from methanotrophic bacteria converting methane, yeast protein from fermentation processes, and microalgae protein (Spirulina, Chlorella) offer sustainable alternatives to fish meal. Production scalability and cost reduction remain challenges.

Processed animal proteins: Rendered poultry byproducts, porcine proteins, and bovine materials can replace fish meal in some species, though regulatory restrictions in certain regions limit use.

Functional Feeds and Nutriceuticals

Immunomodulation: Dietary β-1,3-glucans, vitamin C megadoses, and certain probiotics enhance immune responses, potentially reducing disease incidence and antibiotic use.

Antioxidant supplementation: Vitamin E, selenium, and carotenoids protect against oxidative stress, improving health and product quality. Astaxanthin supplementation produces desirable flesh pigmentation in salmonids.

Gut health promotion: Prebiotic oligosaccharides and probiotic bacteria improve intestinal microbiota, enhance nutrient absorption, and strengthen mucosal immunity.

Precision Nutrition and Feeding Technologies

Nutritional modeling: Computer models predicting nutrient requirements based on fish size, temperature, growth targets, and production goals enable dynamic diet formulation optimizing performance and minimizing waste.

Sensor-based feeding: Underwater cameras, acoustic sensors detecting feeding sounds, and appetite sensors measuring feed consumption enable real-time feeding adjustments, improving FCR by 5-15% compared to fixed schedules.

Metabolomics and genomics: Analysis of metabolic profiles and gene expression patterns in fish tissues reveals nutritional status and requirements more precisely than traditional growth trials, accelerating feed development.

Sustainability Initiatives

Reduced fish meal dependence: Modern salmonid feeds contain <10% fish meal compared to >50% historically, achieved through plant protein optimization and novel alternatives. This reduces pressure on marine fisheries while maintaining performance.

Alternative lipid sources: Terrestrial plant oils (canola, soy, camelina) increasingly replace marine fish oils. Metabolic engineering producing EPA and DHA in oilseed crops promises complete replacement of fish oil in aquafeeds.

Life cycle assessment: Comprehensive environmental analysis of feed ingredients and formulations guides selection toward lower carbon footprint, water use, and land requirements, enhancing aquaculture sustainability.

Genetic-Nutrition Interactions

Nutrigenomics: Understanding how dietary components affect gene expression enables formulation of feeds optimizing metabolic pathways for growth, immunity, or stress tolerance.

Selective breeding for feed efficiency: Genetic selection programs identify families or individuals with superior feed conversion, creating populations requiring less feed for equivalent growth. Genetic gains of 10-15% improvement in FCR per generation have been documented.

Future Directions

Personalized nutrition: Sensor technologies monitoring individual fish appetite, growth, and health may enable tailored feeding strategies maximizing performance of each fish rather than population averages.

Cellular agriculture: Laboratory-cultured fish cells producing protein ingredients without whole-organism production represents long-term possibility, though technical and economic feasibility remains distant.

Artificial intelligence: Machine learning algorithms analyzing multiple data streams (feeding behavior, water quality, growth rates) can predict optimal feeding strategies, detect health problems early, and continuously optimize nutrition programs.

Applied Implications and Future Vision

Nutritional biotechnology offers pathways toward more sustainable, efficient, and profitable aquaculture. Adoption of enzyme supplements, improved plant proteins, and precision feeding technologies provides immediate benefits. Longer-term innovations in novel protein sources and metabolic engineering promise transformation of aquaculture nutrition within coming decades. Progressive farmers integrating scientifically validated nutritional advances while maintaining economic pragmatism position themselves competitively in evolving global aquaculture markets. The convergence of biological understanding, technological capability, and sustainability imperatives creates unprecedented opportunities for innovation in fish nutrition, supporting the continued growth of aquaculture as a crucial component of global food security. Through responsible application of nutritional science and emerging technologies, aquaculture can achieve intensification that is simultaneously productive, profitable, and environmentally sustainable, fulfilling its potential as the most efficient animal protein production system while contributing to human welfare and ecological stewardship.

References

1. Boyd, C. E. (2015). *Pond aquaculture water quality management*. Springer.

2. Chakroff, M. S. (1981). *Aquaculture extension services: case studies* (FAO Fish. Circ. 740). Food and Agriculture Organization. (FAOHome)

3. Coche, A. G., & Van der Wal, H. (1981). *Simple methods for aquaculture: Water for freshwater fish culture* (FAO Train. Ser. 4). Food and Agriculture Organization. (FAOHome)

4. Goddard, S. V. (Ed.). (2006). *Feed management in intensive aquaculture*. Springer.

5. Lekang, O.-I. (2013). *Aquaculture engineering* (2nd ed.). Wiley-Blackwell. (marineinsight.com)

6. Lucas, J. S., & Southgate, P. C. (2012). *Aquaculture: Farming aquatic animals and plants*. Wiley-Blackwell. (University of Hawaii at Hilo)

7. Muir, J., & Roberts, R. J. (Eds.). (2011). *Recent advances in aquaculture: Volume 3*. Wiley-Blackwell. (sanfoundry.com)

8. Noga, E. J. (2010). *Fish disease: Diagnosis and treatment* (2nd ed.). Wiley-Blackwell. (library.viu.ca)

9. Parker, R. (2011). *Aquaculture science* (3rd ed.). Cengage Learning. (marineinsight.com)

10. Soderberg, R. W. (2017). *Aquaculture technology: Flowing water and static water fish culture*. Wiley. (University of Hawaii at Hilo)

11. Stickney, R. R. (2005). *Aquaculture: An introductory text* (3rd ed.). CABI. (marineinsight.com)

12. Tidwell, J. (2019). *Largemouth bass aquaculture*. Wiley-Blackwell. (was.org)

13. Under Gasser, D. (1989). *Handbook of fish diseases*. Wiley. (Goodreads)

14. Van der Wal, H. (1981). *Simple methods for aquaculture: Water for freshwater fish culture* (FAO Train. Ser.). Food and Agriculture Organization. (FAOHome)

15. Wen, P., & Xiang, J. (Eds.). (2021). *Off-shore aquaculture: Spatial planning, zoning and permitting*. FAO/World Bank. (maritimeducation.com)

16. Costa-Pierce, B. (Ed.). (2020). *Ecological aquaculture: The evolution of the blue revolution*. Wiley. (maritimeducation.com)

17. Radosavljevic, S., Venturino, E., Acotto, F., Wang, Q., & Su, J. (2025). *Sustainable intensification of small-scale aquaculture systems depends on the local context and characteristics of producers*. [Preprint]. arXiv. (arXiv)

18. Narisetty, P., Kattamanchi, U. K. R., Nimma, L. A., Karnati, S. R. K., Kore, S.-N. B., Golamari, M., & Nageshreddy, T.

(2025). *AQUA: A large language model for aquaculture & fisheries*. [Preprint]. arXiv. (arXiv)

19. Fisheries of the United States, 2018. (2020). U.S. Department of Commerce, NOAA. (fisheries.noaa.gov)
20. Swindells, P. (2001). *Pond plants and cultivation*. HarperCollins. (Goodreads)
21. Greenberg, P. (2010). *Four Fish: The future of the last wild food*. Penguin Books. (Goodreads)
22. Kurlansky, M. (1997). *Cod: A biography of the fish that changed the world*. Penguin Books. (Goodreads)
23. Jepson, L. (2004). *The super simple guide to common fish diseases*. Wiley. (Goodreads)
24. Helfman, G. (2011). *Fishes: The animal answer guide*. Johns Hopkins University Press. (Goodreads)
25. Hiscock, P. (2007). *Community fish: Sustainable aquaculture for developing countries*. International Centre for Environment & Development. (Goodreads)
26. Garcia, S., & Le Reste, L. (1981). *Life cycles, dynamics, exploitation and management of coastal penaeid shrimp stocks*. FAO Fish. Tech. Pap. 203. (FAOHome)
27. Kapetsky, J. M. (1981). *Some considerations for the management of coastal lagoon and estuarine fisheries*. FAO Fish. Tech. Pap. 218. (FAOHome)
28. Holden, A. V. (1981). *Historical review of EIFAC activities*. EIFAC Occas. Pap. 13. (FAOHome)

29. Coche, A. G. (1985). *Simple methods for aquaculture: Soil and freshwater fish culture* (FAO Train. Ser. 6). Food and Agriculture Organization. (FAOHome)

30. International Centre for Living Aquatic Resources Management. (2006). *Aquaculture, Crustacean and Coldwater Fish Culture*. ICLARM.